GROWING UP IN MUMBAI, INDIA
IN 1940s, '50s AND '60s

Includes stories strangely connected with Diwali, World War I, World War II, India's Independence Movement, Superman, Prithvi Raj Kapoor, New York World's Fair 1964, and more.

SANTOSH KOTHARI

outskirts
press

GROWING UP IN MUMBAI, INDIA IN 1940S, '50S AND '60S
Includes stories strangely connected with Diwali, World War I, World War II, India's Independence Movement, Superman, Prithvi Raj Kapoor, New York World's Fair 1964, and more.
All Rights Reserved.
Copyright © 2021 Santosh Kothari
V2.0 R1.2

The opinions expressed in this manuscript are solely the opinions of the author and do not represent the opinions or thoughts of the publisher. The author has represented and warranted full ownership and/or legal right to publish all the materials in this book.

This book may not be reproduced, transmitted, or stored in whole or in part by any means, including graphic, electronic, or mechanical without the express written consent of the publisher except in the case of brief quotations embodied in critical articles and reviews.

Outskirts Press, Inc.
http://www.outskirtspress.com

ISBN: 978-1-9772-3592-3

Cover Photo © 2021 Santosh Kothari. All rights reserved - used with permission.

Outskirts Press and the "OP" logo are trademarks belonging to Outskirts Press, Inc.

PRINTED IN THE UNITED STATES OF AMERICA

I dedicate this book to my granddaughter Asha and all the children in my extended family who may someday be curious about my life particularly my childhood.

Acknowledgement

I want to thank my wife Sarala, my daughter Sunita, my sons Sundeep and Sanjay, my son-in-law Kapil, and my granddaughter Asha for their support. I am grateful to Sunita and Asha who encouraged me unconditionally. Without their encouragement, this book may not have been completed. I want to thank all in my family, my relatives and my friends who have played such an important role in my life, including those that I may have mentioned in this book under altered names.

Table of Contents

Acknowledgement......v
Preface......i
1. Trip to United States......1
2. My Birth......3
3. Independence of India, World Wars I & II......14
4. My Birthday and Diwali......26
5. My Parents......30
6. Our Home in Sitamau, My Birthplace......39
7. Our Apartment in Mumbai......46
8. Mumbai......57
9. Life in a Mumbai Chawl......61
10. Primary School Days......65
11. High School Days......82
12. Trips to Sitamau......90
13. Trip to Madhya Pradesh......102
14. Romance and Bollywood......110
15. Dream of Becoming a Bollywood Star......124
16. Neighbors......132
17. Our Family Grows Rapidly......137

18. Water Wars 140
19. New Apartment in Andheri 144
20. New Apartment in Santacruz 149
21. Railways 154
22. Education 159
23. Girls 165
24. Science College - St. Xavier's College 168
25. Engineering College - VJTI College 174
26. Trip with Friends 181
27. India Defense Fund and Prithvi Raj Kapoor 197
28. Professional Career Begins 204
29. Student Visa for United States 209
30. Arranged Engagement 212
31. Arranged Marriage 219
32. My Indian Wedding 225
33. Departure for United States 236
34. Superman, Loise Lane, and I 245
35. My Arrival in New York 254
36. Arrival in University of Michigan 259
About the Author 261
Story behind "Growing Up In
Mumbai, India in 1940s, 50s and 60s" 264
Reviews 267

Preface

There is one thing that is for sure—CHANGE. There would be no universe without change. There would be no Big Bang; no gas clouds; no stars; no sun; no earth, no cells, no monkeys, and no human beings. NO change means NO life—Period.

If you agree with Darwin's theory of evolution, changes take place throughout a life, and changes are passed on and carried forward after a life by successors.

If you believe in reincarnation as many in India and some in United States do, change is part of life that results from what you have done in past in this life, as well as what you have done in past lives. This is known as the "Karma" theory. After you die, change will not stop because you will be reborn, and change will continue until Nirvana or Moksha.

If you believe in God, change may take place because of the will of God. For many, change takes place only for a valid reason, which can generally be explained with laws of nature. For others, change takes place because of luck or randomness.

Regardless of the reasons and explanations for it, change has always been there and will always be there. Changes affecting human beings, particularly those caused by human beings, have been gradual during the course of human history. However, they seem to have picked up speed in last 200 years, and the speed of change has accelerated at an exponential rate during my lifetime of seventy-eight years.

Two major events took place when I was an infant and then a child—World War II and the Independence of India. Both of these events and many other events since then have accelerated the pace of change in India, where I was born, and the world at large. This pace has accelerated even further to dizzying levels lately with the introduction of the internet, mobile phones, and other technologies. They may change even more drastically if climate change is not slowed down. Coronavirus Covid-19 has already brought tremendous changes and much more is certain to come, if it is not defeated quickly.

Here are a few examples of how things were during my childhood in India.

There was very little personal transportation—cars, motorbikes, and scooters were rarely used by anyone except the very rich; we did not even have bikes.

Electricity was available only in big cities and only in prosperous areas. When we moved to Bombay at age 2, one or two light bulbs per household was the norm. As I got older, electric fans and then radios were slowly introduced.

Personal visits were the popular method of conducting relations and business. The postal service and telegraphs were the popular methods of communication, if you could not visit. Phones were very scarce and unaffordable.

The Indian population was about a quarter of today's population. There was lot more greenery in cities; a lot more open space; a lot less crowding; a lot less competition; and a lot less greed.

Most people lived in combined families where two to three generations lived together. Often children who were married and had their own children continued to live in the same household. Everything was shared. Before Independence, girls rarely received education past the primary level. After independence, this changed very quickly to middle school level within a decade after the Independence.

Most of us have a desire to leave this world better than we found it. I am no exception. I wrote this book to preserve history in my own personal way for a time, place, and way of life; that is gone.

The period of this book is 1942-1964 and covers the period between my birth and my arrival in the United States. This period was a simpler, more peaceful, more leisurely, and lot less stressful time. That time is gone, forever. I am happy for many changes that have come since then, but I am also sad that the time of my childhood will never return.

Most of the pages in this book describe the way of life during my childhood in Mumbai and other places in India from my perspective and experience. I have not limited myself to my autobiography. I have added more information for context and to share my views on many things. Here is brief description of what you may discover. For example:

- In Chapter 2, you will be able to experience how traveling in Indian train was.
- In chapter 3, you will be able to learn something new in my description of World War I, World War II and Indian Independence struggle.
- In chapter 4, you will know more about the biggest holiday in India, Diwali.
- In Chapter 5, For my family and friends, and to help fully

understand me, I have preserved information on my parents, which may be lost soon.

- In chapter 6, 12 and 13, you will be transported to a village life in my grandparent's days. You will also get introduced to uniquely Indian Fruits and role of animals in village life.
- In chapters 7, 8, 9 and 18, you will get a glimpse of life in Bombay Chawl that hundreds of millions live in India and great part of rest of the world.
- In chapter 10, in addition to my childhood while I was in primary school, you will also be able to enjoy reading about many Indian sports, that rest of the world may not know, exist.
- In chapter 11, you will see how high school education was in a newly independent country with greatest diversity in the world.
- In chapter 13, I have included a section on an Indian regional food.
- In chapter 14 and 15, you will discover how romance and Bollywood are inseparable in India.
- In chapter 16, I describe and recommend exposure to diversity in religion, region, race and other differences to help children grow up trusting "different" people.
- In chapter 17, you will discover how fast and large our family grew because family planning was non-existent.
- In Chapter 19, a personal tragic accident describes unsafe conditions that need more attention in India.
- In chapter 21, A large number of people in mega cities in India use local railways for daily commute. You will get to experience what they experience.
- In Chapter 22, you will get a brief description of Indian caste system and how India has made great progress against discrimination.

- In Chapter 23, "Boy meets Girl" was and continues to be discouraged in India.
- In Chapter 24 and 25 learn how a good education, obtained despite difficult circumstances, has helped many Indians to succeed in india and in United States despite the disadvantages of a foreign language, brown skin and starting with nothing.
- In Chapter 26, join me as I take you on a trip of twelve locations in North part of India on a seventeen days trip covering four thousand miles by Indian trains, along with a detailed history of these places.
- In Chapter 27, read about author's success, at age 19, in getting a great Icon of Indian Film Industry, Prithvi Raj Kapoor, to make an appearance at a fund-raising event at his college at no cost.
- In Chapter 30, 31 and 32, I provide details of arranging marriage which method is still prevalent in India. The wedding with Hindu marriage rituals and their significance is detailed. These rituals may hold the secret of why very few Indian marriages result in divorce.
- Chapter 33 describes chaotic scene at Author's home because about 100 family members arrived to see him off.
- Chapter 34 is a "Must Read" so you will know why such a strange title.
- Chapter 35, 36 and 37 describe eventful first week of Author's arrival in United States.

CHAPTER 1

Trip to United States

I WAS SITTING in the plane looking down at clouds below. It was August 17, 1964. After about 9 months of crossing bureaucratic red tape, I had obtained a student visa to go to the University of Michigan, USA, to study for a master's degree in Civil Engineering. My boss at my last job in India told me that if I came back to work for them after completing my studies, I would be getting a raise to Rs 800 per month; an increase from my last salary of Rs 350 per month. So, when I decided to go to the United States, it was a no-brainer. I would easily recover cost of my travel and education by working 18 months that I would be allowed to work on a student visa to obtain practical training. Plus, I would be making more money for the rest of my life. There was no cost and no risk to me.

I had no idea at that time that I would never go to go back to live in India; that I would be at the top of my class at the University of Michigan; that I would get a chance to perform songs and dances in front of thousands of people at University of Michigan as an international student; that I would succeed professionally as a structural engineer; that I would start an engineering company; that I would design over a thousand multi-million dollar industrial and commercial projects; that I would help transform a group of Jain families into an organized Jain Society of Greater Atlanta and become its President;

that I would lead the society to acquire land for the Jain Center; that I would help build the Jain Center and Jain Temple; that I would help bring the JAINA Convention to Atlanta; that I would develop many multi-million dollar commercial properties and own them; that I would be able to produce a selfie of my Bollywood remix video song and post it on Youtube.com for the entire world to see; that I would be a proud parent of three lovely children—a daughter who would become a doctor, a son who would become a lawyer, and my youngest son who would become an information technology consultant and who would start a charity to help young kids with long term ailments; that I would have only one grandchild—a daughter; and last but not least that I would be writing this book 55 years later.

Author's flight to New York on August 17, 1964 was in Air-India Boeing 707.

I did not know any of this, but I was excited. I just wanted to soak in every moment as I looked forward to my educational trip to the USA. I did not want to sleep, but the vibration and constant humming of the engine and vibration of the plane made me drowsy. Suddenly I was seeing Sitamau, the place where I was born.

CHAPTER 2

My Birth

THE YEAR WAS 1942, five years before India gained its independence from British rule of almost two hundred years. A young woman, named Subhadra Jainendra Kothari, my mother, whom we called Baiji, aged 23, was traveling from Bombay to Ratlam, Madhya Pradesh, where she would change trains to go to Mandsaur, from where she would take a bus to reach Sitamau. She was seven months pregnant. Even though she was accompanied by her husband, my father, Jainendra Kumar Kothari, who we called Babuji, she did not really want to make this long tiring trip. But her mother-in-law, my grandmother, who we called Dadiji, had insisted that my mother go there so that she would be better taken care of. In Bombay, renamed Mumbai in 1995, at Baiji's home, there were no elderly ladies who knew how to deliver a baby and provide pre-natal or post-natal medical services. Therefore, Baiji reluctantly agreed to make this trip.

When India obtained its independence from British rule, it also inherited a corrupt bureaucracy and a poor infrastructure. Almost all new roads that were built by the British government had the primary purpose of allowing the British government access to various parts of India so they could more effectively govern. Train system, which did not exist until the British government built the system, was similarly greatly influenced by British interests.

GROWING UP IN MUMBAI, INDIA IN 1940S, '50S AND '60S

When I was born, and even now seventy-seven years later, if you wanted to travel long distances, train was usually the best mode of transportation, and most often the only practical choice for most people with a limited budget. You could either travel in reserved compartments or non-reserved compartments. You could either travel first class, which had leather-covered cushioned seats, or second class with glass fiber reinforced plastic (renamed from formerly third class and improved from wooden benches). First class fares were usually five to ten times the third-class fares, so first class was beyond reach for my family. In order to reserve a seat, you had to go within the first hour or so from the time that reservations were open for sale; otherwise, they would all be gone. The reason for such fast sales was that many of the seats were sold to specially privileged government employees, politicians, or someone who was good at bribing railway employees. If you did obtain a reserved seat, you were assured a place in the special reserved compartment and your journey would be fairly comfortable.

Travel away from home, even if it was just for a few days, involved a ritual that may be unique to India. India, for most of its history of more than 5,000 years, has been one of the most populous regions in the world. When people are crowded together, most people hate the crowded condition. Even so, crowded conditions brought people in India closer to each other. As soon as Baiji's trip was announced, many of her friends and relatives started visiting to see her one last time before she left for a trip that might last a few months.

The number of people visiting her kept increasing as the day of the trip approached. Everyone who was visiting from out of town stayed with us since the hotels were not affordable or conveniently located; or more likely because everyone in India loves to save money. Local people came and stayed for part of the day and went home at night.

Everyone wanted to help. They helped with last minute shopping. They helped with packing her bags, meals, bed sheets, blankets, a pillow for the overnight train trip, and anything else that they thought she might need for her trip.

During this period, our house was a noisy place. Indians love to talk. There are many reasons for this. First, everyone is constantly trying to prove how smart they are, so they are constantly correcting each other and explaining their deep understanding of every subject that comes up. Another reason is that Indians who grow up in large, noisy families cannot stand silence. Conversations help them forget their grief, their problems, and their worries.

There were only two rooms—the living room and kitchen. So, all of the men, most of them dressed in the popular white clothes, piled up in living room, some on the few chairs that were available, and the rest of them on floor, cross-legged or with their feet stretched out. Women, most of them dressed in bright and dark colored saris, sat on the floor cross-legged in the kitchen. If they were out of sight of men, they had the freedom to uncover their head. When they could, they were very happy to do so.

Women were always cooking or preparing things to cook later. There were many rounds of tea, snacks, and meals. Rooms were filled with inviting sounds made by aluminum, brass, or stainless-steel plates/thalis, bowls/katoris, water glasses, and serving utensils/tapelis, as food and drinks were served, re-filled, and taken away and as they were laid on the colorful tiled floor where everybody sat to eat. Varieties of aromas of delicious food floated in the air all day long. My mouth waters just thinking about it.

Three hours before departure time, Babuji and Baiji got ready to go to train station. Most of the people who were at home, particularly those who had traveled from out of town, also piled up in several tangas/one horse carts or took buses and local trains to reach the train station.

GROWING UP IN MUMBAI, INDIA IN 1940S, '50S AND '60S

Now you may wonder why all these people, about thirty in all, were going to the train station, when only Babuji and Baiji were traveling. I can't fully explain this. All I can say is that they wanted to be with Babuji and Baiji, not only up until the very last moment when the train would start to leave the station, but even after, until they could no longer see the train departing.

The train was the Western Railway leaving from Bombay heading to Delhi, with Ratlam roughly halfway, approximately 400 miles from Bombay. The train was going to leave from the Bombay Central Station. The tanga trip from our home in Mahim was only a half hour long but getting to Bombay Central Station way ahead of schedule was a good idea. Most Indians had acquired such a habit. For one thing, you were not sure about the traffic. More importantly, you wanted to get there before the departing train arrived so that you could rush into the departing train when it arrived. You did so that you could occupy your seat and keep it protected.

Since Baiji and Babuji had reservations, you would think that they could get on the train even at the last minute, and there wouldn't be a problem. But unfortunately, such was not the case then, and it is not the case today.

The reason for this is that many people who do not have reservations also pile into the reserved compartment. Then when the Ticket Checker (TC) comes to check everyone's reservation, those who do not have a reservation are asked to step aside for a private meeting with the TC. In this private meeting, the person begs and explains why they are in the reserved compartment without a reservation. The TCs are very understanding, particularly if a bribe is given. The TC quotes a price which includes the fare and bribe. Final amount is negotiated and paid. The TC comes back with the passenger and requests the reserved seat holders to squeeze and accommodate the extra passengers. Everyone goes along with the TC. Therefore, it is not uncommon

MY BIRTH

to find four, or sometimes even five people, squeezed onto the bench which is designed to seat three.

In addition to these non-reserved people who bribe and are accommodated by TC, there are many passengers who would not be assigned seats, but who would still be allowed to ride in the compartment standing or sitting on the floor, most of whom paid a smaller bribe.

At night, the space between seats and passageway to exits and restrooms are filled up by these non-reserved passengers. Those who have reservations get to sleep on one of three boards which are thirty inches apart vertically when opened up. These boards are folded out of the way during the day. If you are in top two boards, you use a ladder at the end, or jump onto and off the board. You sleep on a bed sheet or a thin comforter and cover yourself with another bed sheet to keep warm and to keep mosquitoes away. Most people do not carry a pillow, so they use their shoes, a bag, or a purse as a pillow. Seats are assigned without regard to gender, but most men out of courtesy give up their lowest seat to any woman who is generally respected like a sister or a mother.

Most of the time, people preferred to start their journey at night so that they reach their destination next morning. Sleeping at night makes the journey go so much faster, and it saves time. Babuji and Biaiji left Bombay about 8 p.m. and were scheduled to arrive in Ratlam at about 2 p.m. the next day. It was an eighteen-hour journey to Ratlam. The average train speed was thirty to thirty-five miles per hour. When you include train stops, the average speed dropped to an average of twenty-five miles per hour.

Until the train takes off, you are going to be very busy. First you make sure that everything that you expected to carry has been delivered to the train. This is a serious concern with so many people, including coolies, carrying your stuff in their zeal to help, because everyone wanted to carry the bags. And those who did not get a chance initially insisted on relieving a person who was carrying a bag even if he was doing fine. Thus, each item of baggage could have been handled by several persons. It is hard to concentrate, with all the people who come to see, but you try. Everyone who comes to see you off are trying to get a last word of advice in. In addition, they also want to give you all the messages they want you to convey to people you are going to see in your trip. Even though you enter the train at least an hour before departure, time flies. Before you know, the time for the train to depart has come. The train conductor gives a whistle to warn passengers that train is going to start moving and they'd better get on board. Also, to warn everyone on platform that they should keep clear of the moving train. He waves a green flag to the engineer. The engineer starts the train. All the people who came inside the train now start to get off while the train is pulling out. All the passengers who were still on the platform, chatting with the people who came to see them off, get into the train just as train picks up speed. Soon the train leaves the platform. All the persons who had come to see the travelers off, start heading for exits in groups.

Passengers settle down in their seats, you make acquaintances with all the people near you and eat snacks or meals. Baiji had come fully prepared. She pulled out a four-compartment tiffin, which had more food than Babuji and Biji could consume. She and Babuji ate and also shared and exchanged food with strangers in the same area. They were already beginning to make friends for the trip. For the next 18 hours of the trip, they discussed politics, children, relatives, their medical conditions, and any and all subjects that they thought about. What is striking is that they shared details of their life that they might not share with people that they meet every day in their lives. They could do that with strangers because there were no long-term repercussions. Babuji played cards with them. Teen patti/three card poker was popular. Baiji shared recipes. They told them stories and jokes, and others did the same. While Babuji did not sing and Baiji only sang

MY BIRTH

Bhajans in private settings, there were some on the train, like half the population of India, who loved to see film songs accompanied by tapping tabla-like on a bench. Some of them were good singers, others were not, but they all got a chance. When someone was singing, it was ok for others to join chorus style. If no one was volunteering, then a game called Antakshari, meaning start with last letter, was played. Now good singing was not as important as a good memory.

Pretty soon it was time to sleep. Baiji went to sleep. Babuji just looked out the windows and enjoyed an ever-changing night scenery with scant light from homes and the moon. Some of the people even read in the dim light of the train, which flickered a lot, as the voltage generated by the steam engine fluctuated. Babuji slept on and off. Soon it was 6:00 a.m., and day was breaking open.

When they arrived at the next station, there was a lot of noise. They could hear vendors promoting and selling chai/tea, Samosa, Batavada, Bhajia, Ganthia, Chevda, Jalebi, and many sweets, snacks, and hot and cold drinks. Pre-packaged food items were not popular at that time and were also generally prohibitively expensive. Most Indians ignore health hazards of fresh foods. It was hard to resist steaming fried food, inviting sweets, or a steaming cup of tea. This was even more so when you saw others enjoying these delicious foods or drinks. Vendors were walking by the windows of the train and shouting slogans like "Garam Chai Garam Chai/hot chai" twice in a singing voice and selling these items, exchanging items for money through the windows. Tea was sold in disposable clay mugs, and freshly made snacks were sold wrapped in old newspapers. Indians don't seem to care that toxic printing ink leach into their food. Some vendors sold fruits from a basket. You could see that basket attracted a lot of flies and mosquitoes. But Indians have just learned to ignore any danger from exposure to flies and mosquitoes. They have to. Some people got out of the train. They washed their faces, swished their mouth as a short cut to brushing from consumed drinks and snacks. They bought

GROWING UP IN MUMBAI, INDIA IN 1940S, '50S AND '60S

newspapers and magazines. They also refilled their water containers from a water tank. When the train started, people lined up to use the toilet on the train. Using the toilet while train was stationary was prohibited because human waste was delivered spontaneously to the railroad track below through an open bottom toilet.

The long ride in the train was fine, except for the constant jerk and noise made by the wheels as they crossed the expansion joint between uneven and poorly-maintained rails. And then there was the coal dust and coal ashes from the coal that was burned in boilers to power the steam engines pulling the train. Coal dust particles are hard and sharp. They hurt and irritated your eyes as they flew at you at high speed through the open window of the train. Coal dust went into your nose, your ears, and your hair. The coal dust made your face and your clothes black. When you finished your travel, the first thing you wanted to do was to take a nice and long cold water bath and scrub yourself with lot of soap.

Indian Railway's Coal-Fired Steam Locomotive

MY BIRTH

After a while, the train trip was starting to get monotonous. The same scenery was going by as train slowly moved from station to station. Most of the landscape consisted of brown uncovered ground, broken by a few small trees and some weeds. You saw small huts, cows, bullocks, stray dogs, and some monkeys besides people walking on unpaved roads. There were also a few men and women going through their morning routines including brushing their teeth and going to the toilet in the open.

Finally, Baiji and Babuji reached their destination in Ratlam. The train was going to stop in Ratlam, so there was no hurry. But a less hectic version of what happened in Bombay happened in reverse. There were a lot of people who had come to the station to receive Baiji and Babuji. Most patiently waited outside the train. Coolies were hired to carry luggage, including the smallest bags. Some of the people who had come to receive Baiji and Babuji rushed into the train to greet them and welcome them. They were so excited; they could not wait until Baiji and Babuji got home or at least got off the train. They wanted to know how the trip was, how everybody back in Bombay was, and so on and so forth. There were lots of smiles and laughter. Joy of seeing someone you love after a long time is indescribable. Everybody piled into a few tangas that had been already reserved, and they reached my Bade Mamaji's home.

Baiji and Babuji spent a few days before continuing to their final destination Babuji's home in Sitamau. From Ratlam, they had about a three hours' train ride to Mandsaur, and then an hour ride in the state-run bus to Sitamau. The bus rides in state-run buses over roads that were mostly dirt, with a little bit of gravel, with a lot of uneven surface and potholes was slow, dusty, bumpy, and shaky. But Babuji and Baiji were excitedly looking forward to seeing Dadiji, Dadaji, and my uncles/Kaka Saheb, and my aunts/Bhua Saheb, and their families. After a long trip from Bombay, Baiji and Babuji finally arrived in Sitamau.

Babuji went back to Bombay after a few days to resume his job. Baiji stayed about four months in Sitamau, roughly two months before and two months after the delivery. It is customary for Indian mothers to be fed twice the normal quantity during pregnancy. It is believed that mothers have to be fed not only for herself, but also for the soon-to-be-born child. Food was required to be nutritious, which according to Indian transition meant it had to be rich in calories and fat. Baiji was asked to rest most of the time. She was not allowed to help around the house or exert herself in any way for fear that any carelessness may result in miscarriage. Baiji was amused at all the fuss being made. She was already experienced at having babies. She had already delivered four children—two boys and two girls. Nonetheless, she did as she was told.

On November 8, 1942, and on the day in the month of Kartika and on the day of Amavasya, night of the new moon, on the day of one of the holiest festivals, Diwali, Baiji gave birth to a healthy boy. I began my life in Sitamau.

I was born at a time of tremendous change when India got independence from foreign rulers who controlled large parts of India for 800 years. I was lucky to escape World War I and World War II. Misery of these two world wars was beyond comparison to any other tragedy that world has seen before, and I hope the world never sees again.

My parents named me Santosh. Santosh means contentment. My name was intended to inspire me to always be happy and make everyone else happy. I am not sure how much I have succeeded. But I will keep trying.

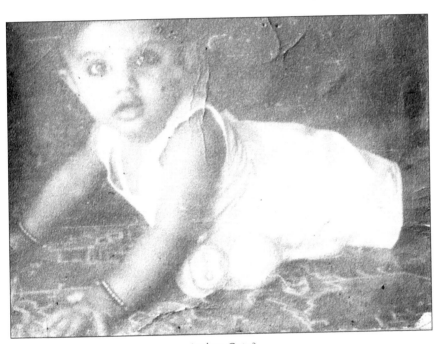

Author. Cute?

CHAPTER 3

Independence of India, World Wars I & II

Indian Independence Movement:
THE DEFEAT OF Prithvi Raj Chauhan in 1200 A.D. started an almost 800 years long period in which great parts of India were controlled by foreign powers. First, Muslim for nearly five hundred years. Then the British for nearly two hundred years, along with Dutch, Danish, French, and Portuguese to a lesser extent during the same period.

Desire for independence in various forms and to various degrees always existed during this period. But the intensity of organized efforts increased significantly after 1857 until India got independence on August 15, 1947.

In the 1700s, efforts were made by many who fought the British for various injustices: Puli Thevar of the Tamil Nadu region in 1760s; Syed Mir Nisar Ali Titumir in the Bengal region in 1820s; A more organized effort was made by Hyder Ali and Tipu Sultan of Mysore, Maratha Kings and Nizam of Hyderabad during 1770-1800; Jagannath Gajapati Nararyan Deo of Odisha in 1760s; Kerala Varma Pazhissi Raja from 1780-1800; Rani Velu Nachiyar from 1760-1790; Veerapandiya Kattabomman from 1790-1800; and Dheeran

INDEPENDENCE OF INDIA, WORLD WARS I & II

Chinnamalai from 1800-1805. In 1804, Jayee Rajguru, followed by Bakshi Jagabandhu in 1806, rebelled in an armed rebellion in Odisha known as the Paik Rebellion.

The Indian rebellion of 1857, which may be considered the beginning of the final push for independence, was caused by various injustices against the soldiers and various rulers. It was precipitated by rumored use of tallow from cows which is offensive to Hindus, and lard from pigs which is offensive to Muslims, in the newly introduced rifle cartridge. Soldiers had to bite the cartridges with their teeth before loading. Mangal Pandey killed a British officer in protest. When he was captured and sentenced to death, the protest spread, lasting from 1857 to 1859. After Rani Lakshmibai of Jhansi was killed in 1858, the rebellion was by and large crushed. As a result of the rebellion, the British government took control of the rule in India from the East India Trading Company and the British Raj began, which lasted until 1947.

Even though the British made a number of changes to placate the Indians, discontent among the population led to continued organized efforts to fight for more rights and justice.

Dadabhai Naoroji formed the East India Association in 1867. Surendranath Banerjee founded the Indian National Association in 1876. The Indian National Congress was formed in 1885, first as a debating society in which Indian elites such as lawyers, educators, and journalists participated. To begin with, Congress expressed loyalty to British rule and primarily worked to get small concessions. Though not very effective, it represented the first efforts to form a group representing all Indians, and also to form the idea of one nation to replace a collection of more than 500 princely states.

Further progress was made by many individuals and their organizations, such as Swami Dayanand Saraswati (founded Arya Samaj), Ram Manohar Roy (founded Bramho Samaj), Swami Vivekanand,

GROWING UP IN MUMBAI, INDIA IN 1940S, '50S AND '60S

Ramakrishna, Sri Aurobindo, Chidambaram Pillai, Subramanya Bharathy, Bankim Chandra Chatterji, Sir Syed Ahmed Khan (founded Aligarh Muslim University), Rabindranath, and sister Nivedita.

Until 1907, congress members considered themselves loyal British citizens, but wanted a greater role for Indians in the governance of India. Bal Gangadhar Tilak, on the other hand, demanded Swaraj as Indians' birthright. Lala Lajpat Roy and Bipin Chandra Pal supported him. They formed the radical wing of the congress party. They wanted to overthrow the British government and wanted to abandon all things British. Moderates lead by Dadabhai Naoroji and Gopal Krishna Gokhale wanted reform within British Rule. Tilak, backed by Bipin Chandra Pal and Lala Lajpat Rai, raised the demand for independence in Maharashtra, Bengal and Punjab, and were thrown out of congress and were arrested. Because of the split, Congress lost some credibility. The Muslim League separated from congress.

In 1905, Bengal was divided into two provinces—East Bengal with Muslim majority and West Bengal with Hindu majority. Indians viewed this partition as an attempt by the British to disrupt the growing national movement and to divide Hindus and Muslims. In protest, Congress launched the Swadeshi movement, boycotting British products. Activities such as newspaper articles challenging British, and violent actions for independence against British increased, and many Indians were punished with imprisonment and death sentences. In 1916, Muhammad Ali Jinnah joined the Indian National Congress. Due to many factors, including contributions of Indians to World War I, on Behalf of the British Government, the British introduced a dual mode of administration, or diarchy, in which both elected Indian legislatures and appointed British officials shared power. Portfolios such as agriculture, local government, health, education, and public works were handed over to Indians. Portfolios such as finance, taxation, and maintaining law and order were retained by British administrators.

INDEPENDENCE OF INDIA, WORLD WARS I & II

Gandhi's arrival in 1915 from South Africa was going to give a new direction to efforts to gain independence. Until then, all the efforts were limited in geographical reach and also limited in their objectives. Even though Gandhi also started with a limited objective of obtaining justice, in just a few years his objective widened to include independence for all Indians.

Gandhi was a leader of the Indian Nationalist Movement in South Africa. He was a vocal opponent of basic discrimination, abusive labor treatment, and police suppressive methods. He perfected a new method unknown to the world, called Satyagrah—insistence on truth. With this method he was able to get legislative reform and the release of Indian political prisoners in South Africa. His methods for himself and for his followers were strictly non-violent protests, boycotts, marches, hunger strikes, etc. He was greatly inspired in his method of truth and non-violence, by a young Jain monk named Shrimad Rajchandra, who wrote long letters to answer Gandhi's question on religion and philosophy.

After the end of World War I in 1918, a commission was set up to look into the wartime conspiracies by nationalists and to recommend measures for post-war period. The commission recommended the extension of extraordinary wartime powers to quell sedition by silencing the press, detaining political activists without trial, and arresting individuals suspected of sedition or treason without a warrant. Police used these methods widely and indiscriminately. A nationwide strike marked the beginning of widespread popular discontent.

The British attacked the demonstrators, which culminated in 1919 in the Jallianwala Bagh massacre in Amritsar, Punjab. After blocking the only avenue for escape, General Dyer ordered his soldiers to fire into an unarmed and unsuspecting peaceful demonstrator crowd of 15,000 men, women, and children. He was going to teach the Indians a lesson. Dyer was asked to retire and returned to Britain.

However, to the Indians great disappointment, he was held as a hero upon his return. Indian Nationalists knew then that nothing short of complete self-rule would suffice.

Gandhi's campaign went into high gear. Gandhi started the Non-cooperation Movement. Gandhi convinced other leaders to start the Non-cooperation Movement in support of self-rule. All leaders urged the people of India to use Khadi, a home-spun cotton material, as an alternative to British-made imported textile. They urged the people to boycott British schools, British courts, government employment, refuse to pay taxes, and to give up British titles and honors. There was widespread popular support for such actions which resulted in disorder. The movement was called off because of fear of angry and violent reactions by Indians. The party, which was limited to a few elites primarily involved in discussions, changed to a party where any Indian could be a member for a token fee. The party became a vehicle for mass participation.

Gandhi was sentenced to six years which was reduced to two years for his role in the disruption. Upon his release in 1922, he set up the Sabarmati Ashram in Ahmedabad, Gujarat. He established a newspaper, *Young India*. He introduced a number of reforms in the Hindu Society for the benefit of untouchables, backward classes, and rural poor Indians. He also groomed many Indian leaders who supported his views. Prominent among them are C. Rajagopalachari, Jawaharlal Nehru, and Vallabhbhai Patel. Leaders who did not agree with his methods also emerged, such as Subhas Chandra Bose who formed the Indian National Army. Many national and regional political parties also emerged. Many writers, poets, and women played important roles in the fight for independence.

In 1928, a drafting committee under Motilal Nehru was appointed to draw up a constitution for India. In 1929, the Indian National Congress adopted the objective of complete self-rule. January 26, 1930 was ob-

INDEPENDENCE OF INDIA, WORLD WARS I & II

served throughout the country as Purna-Swaraj/Complete Self-Rule Day. India was united in observing this day with honor and pride. In 1930, Gandhi led a march to produce salt in Dandi, Gujrat as an act of nonviolent civil disobedience. The British had taxed production of salt from seawater and used that tax to prevent salt production. The march got worldwide attention and expanded the nonviolent civil disobedience movement.

From 1931 to 1935, Congress and the government were locked in both conflict and negotiations. In the meantime, rifts between Congress and the Muslim League widened. The Muslim League questioned Congress's claim to represent all Indians. Congress questioned the Muslim league's claim to represent all Muslims.

In 1939, the Muslim League passed a resolution, known as Lahore Resolution, which demanded division of India into two nations—one Muslim and another Hindu. In opposition, many Muslims participated in the All India Azad Muslim Conference in Delhi in 1940, supporting united India. Attendance in the Delhi conference, which supported united India, was five times as big as that in the Lahore conference.

From 1939 to 1945, a number of events took place. They did not necessarily pull all in the same direction. Even so, they all had one thing in common—Indians were tired of British rule and the British needed to leave India.

In the 1937 provincial elections, Congress did extremely well. Desire for self-rule was very evident from this vote. When World War II started, the Indian Viceroy declared that India would fight on the British side without consulting the Indian representatives.

The entire congress leadership resigned from provincial and local governments. In 1942, Congress launched the Quit India movement.

GROWING UP IN MUMBAI, INDIA IN 1940S, '50S AND '60S

The British government cracked down and arrested tens of thousands of congress leaders, including all the main national leaders and provincial figures, and detained them until the end of World War II was in sight in 1945.

There were number of independent revolutionary movements, including the Azad Hind Movement under Netaji Subhas Chandra Bose, who broke away from congress to do so, and the common man's revolutionary movement under Lal Bahadur Shashtri.

In 1946, Indian sailors revolted on board ships and also on shore establishments at Bombay Harbor. The struggle spread and found support throughout India. Indian people joined the twenty thousand sailors and were involved in agitations, mass strikes, and demonstrations. It was obvious that if India did not get independence soon, chaos would follow.

World War II accelerated the independence of India and many British and non-British colonies. Modern colonialism involved many European empires starting in the 15th century until the end of the World War II. Primary among them were British, Belgian, Danish, Dutch, French, Hungarian, Norwegian, Portuguese, Russian, Spanish, and Swedish empires. In addition, Japan, the Turkish/Ottoman Empire, the United States, China, Germany, Italy, Australia, and New Zealand also became colonial powers. There were additional colonial powers in South America and Africa. Many of these colonies started out to gain trade advantage. But by the 19th century, their purpose included military control and the spread of Christianity.

Before World War II, 85% of the globe was controlled by colonial powers. But by 1975, most colonies had become politically independent. About forty countries got independence in twenty years after the World War II ended. Important reasons for this, besides the local conditions, were the poor economic state of former colonial rulers

INDEPENDENCE OF INDIA, WORLD WARS I & II

after the War and anti-colonial sentiments of the resulting two superpowers— the United States and the Soviet Union.

A decision to grant independence and partition India was made and accepted by the Indian people. Pakistan got its independence on August 14, 1947, followed by India on August 15, 1947. Violent clashes between Hindus, Sikhs, and Muslims followed. Chakravarti Rajagopalachari was the first Indian Governor General. Sardar Vallabhbhai Patel was made Home Minister and organized bringing 565 princely states into the Union of India with his "iron fist in a velvet glove" policies. Dr. Ambedkar headed the constituent Assembly. On January 26, 1950, the constitution was adopted, and the Republic of India was officially proclaimed. Dr. Rajendra Prasad was the first elected President of India.

Jawaharlal Nehru delivering his speech- Tryst with Destiny, on the eve of India's first Independence Day on August 15, 1947.

While Mahatma Gandhi should get a big share of credit for India's independence, India's independence was not won by a single victory in a single battle. It was won by many Indians who fought for a long time and who never gave up despite reversals. Because of this, credit for creation and sustaining of the new democratic independent India belongs to all of its people.

I was two when World War II ended. I was four when India became independent. I was seven when the Republic of India was proclaimed with a constitution guaranteeing total democratic control by Indians for their own destiny. I feel lucky about the timing of my birth.

World War I:

World War I, lasting from 1914 to 1918, was hoped to be "the war to end all wars". This war mobilized 60 million European military personnel. It led to death of 9 million military and 7 million civilians. It led to an influenza pandemic causing death of 50 to 75 million people all around the world in 1918.

World War I began on June 28, 1914, with the assassination of Austrio-Hungarian Heir Ferdinand by Bosnian Serb Yugoslav nationalist Gavrilo Princip. The fight between Austria-Hungary and Serbia began due to this event. War quickly expanded from a fight between two countries into one involving the entire Europe in just a few months. Two opposing coalitions emerged. France, Russia, and Britain formed what came to be known as the Triple Entente. Opposing this was Germany, Austria-Hungary, and Italy, what came to be known as the Triple Alliance, with Germany in the leadership position. In another month, the war expanded to include Japan on Entente's side and the Ottoman Empire on the Alliance's side. Colonial empires were quickly sucked into the war, spreading the war to Europe, Africa, Middle East, Pacific Islands, China, the Indian Ocean, and the North and South Atlantic Ocean. As the Triple Entente expanded, it came to be known as the Alliance Powers. The expanded opposing German side came to be known as the Central Powers. Until 1917, the United States stayed out of the war except for supplying material to the Alliance Powers. But the sinking of an American merchant ship and efforts to provoke Mexico into war with the United States lead to the United

INDEPENDENCE OF INDIA, WORLD WARS I & II

States troop involvement which quickly grew to two million troops during the war.

The war seemed to wind down over the 3-year period with ups and down for both sides. The final conclusion came in 1918, starting with Bulgaria, which was forced to sign an armistice in September 1918. The Ottoman Empire did so in October 1918. Austria-Hungary did so in November 1918. The Germans signed an armistice on November 11, 1918, which marked the end of World War I.

The Big Four—Britain, France, the United States, and Italy won the war. Austro-Hungarian, Germany, the Ottoman Empire, and the Russian Empire ceased to exist, and a number of new states were created. World War I was a great turning point in the cultural, economic, and social climate of the world. The League of Nations was created. It seemed that the world would never see another war like this ever again. However, only twenty years later, World War II followed.

While role of Indians and the Indian Army is hardly mentioned, the Indian Army contributed over a million troops to European, Mediterranean and Middle East theatres of war. Approximately 75,000 Indian soldiers died, and an almost an equal number of Indian soldiers were wounded.

World War II:

World War II was a global war that lasted from 1939 to 1945. The war involved more than 100 million people from most countries of the world. This is at a time when the world population, at 2.5 billion, was about a third of today's population. It was the deadliest conflict the world has ever seen. It resulted in 50 to 100 million deaths, and a large portion was civilian due to direct actions of the war and to an even greater extent from starvation and diseases.

GROWING UP IN MUMBAI, INDIA IN 1940S, '50S AND '60S

Two opposing alliances were formed. The Allies on one hand and the Axis on the other.

The beginning of war was marked by Germany's invasion of Poland in September of 1939. France and the United Kingdom declared war on Germany. Japan, which aimed to dominate Asia and the pacific, was already in an undeclared war with China by 1937.

During the early part of this world war, Germany conquered or controlled most of continental Europe. Germany had formed the Axis Alliance with Italy and Japan. Between Germany and the Soviet Union, under a treaty in 1939, they partitioned and annexed territories of their European neighbors, Poland, Finland, Romania, and the Baltic States.

After the fall of France in 1940, campaigns that started in North and East Africa continued between the Axis Powers and the British Empire. In 1941, the Axis powers launched an invasion of the Soviet Union. In December 1941, Japan launched a surprise attack on Pearl Harbor in Hawaii, United States, and on European Colonies in the Pacific. The U.S. declared war against Japan and was supported by Great Britain. The Axis Power and Japan declared war against U.S. Japan had great early success. However, the Axis advance in the Pacific was halted in 1942 after their defeat in the Battle of Midway. Germany and Italy were defeated in North Africa and then decisively in Stalingrad in Soviet Union. The Axis had series of setbacks in 1943.

In 1944, the Alliance invaded German-occupied France, while the Soviet Union regained most of its territorial losses. During 1944 and 1945, Japan suffered major losses in Asia and the Pacific Islands. The Axis was weakened and the Allies and the Soviet Union invaded Germany, leading to the capture of Berlin, the suicide of Adolf Hitler, and finally, an unconditional German surrender on May 8, 1945. When Japan refused to surrender, the United States dropped atomic

INDEPENDENCE OF INDIA, WORLD WARS I & II

bombs on the Japanese cities of Hiroshima and Nagasaki. This was the first and only time atomic bombs have been used in the world. A few days later, on August 15, 1945, Japan surrendered. World War II ended, and war crimes trials against the Germans and Japanese concluded the war.

I was less than three at that time. I was lucky that a relatively peaceful seventy years followed after this. Nuclear threat that appeared imminent at my birth has so far not materialized. The United Nations was formed. The Soviet Union and the United States emerged as two superpowers and engaged in a cold war which still continues today. With Europe weakened greatly, colonial empires collapsed. Most of Asia and Africa, including the Indian subcontinent, emerged as independent nations. Philippines gained independence from the United States in 1946.

The United States emerged as the most powerful country with expanded territory and power. Alaska, which had been purchased in 1867, became a state of United States in 1959. Hawaii, which is made up of hundreds of islands stretched over 1500 miles, became a state of the United States in 1959. Puerto Rico, American Samoa, Guam, the U.S. Virgin Islands, and many other territories throughout the world were retained by the United States to serve its military goals. The U.S. retained ownership of over 800 military overseas bases around the world, such that an American president can fly all around the world without leaving American soil.

I consider myself very lucky that I was born after all the hard struggle to gain independence and all the misery caused by two world wars was behind us. I was born into a long period lasting my life time that turned out to be more peaceful and more prosperous.

CHAPTER 4

My Birthday and Diwali

SITAMAU WAS A PRINCELY state under British rule in the Malwa Region, in the Mandsaur District, in the state of Madhya Pradesh, India. My birth was not recorded in any public documents. I do not have a birth certificate. My school leaving certificate helped me prove that I exist. I was able to obtain passports which have served as my proxy birth certificate.

While I cannot take credit for choosing the date of my birth, I do feel very lucky to have been born on the day of Diwali. Everyone who finds out that I was born on Diwali expresses surprise and happiness at how lucky I am. I am sure they also feel a little jealous. It is difficult to forget my birthday. There is no excuse. And even if it is forgotten, it is very easy to celebrate it at the last minute because the important and time-consuming work of preparing good food and gathering friends and family is already done.

My birthday, Diwali, falls on a varying day in the Gregorian calendar currently used by everybody. All the important Hindu festivals such as Holi, Maha Shivaratri, Vaisakhi, Raksha Bandhan, Krishna Janmashtami, Durga Puja, Ram Navami, and Diwali, are generally set by the Vikram calendar, which begins in 57 B.C. This calendar is a lunisolar calendar. Each month is 28 days and divided into two parts

MY BIRTHDAY AND DIWALI

of 14 days each—shad/rising moon and vad/diminishing moon. Extra days and an extra month are added in a complicated mathematical way to bring the Vikram calendar year in line with actual lunar year of 354 days, and also the actual solar year of 365 days, 6 hours, and 12 minutes, approximately.

Diwali is the greatest day of celebration for most everyone in India, for most of the Indian American families in the United States, and all the Indians living in every country on every continent. Indians are everywhere. The larger the Indian population, the larger is my birthday celebration. It is a very important day for over eighty-five percent of the Indian population who are Hindus, Jains, and Sikhs. But many Indians, including Buddhist, Jews, Christians, and Muslims, and other religions also celebrate it all over the world. It is difficult to escape the spirit of Diwali if you are Indian or if you are anywhere near an Indian.

Diwali is celebrated with oil lamps made of clay. Special prayers to goddess of wealth Laxmi and god of prosperity Lord Ganesh are important part of this festival of lights.

Diwali is also known as the "Festival of Lights". The reason is that it is celebrated with lots of Diyas, a clay or porcelain cup with oil in which a cotton wick is placed and lighted. It will usually burn for hours, even without refilling.

There are many stories connected with Diwali day. One of the most popular among Hindus is that the festival celebrates the triumphant return of Lord Rama after his victory over the evil King Ravana about 3000 years ago. It is also interpreted as a victory of light over darkness, a victory of knowledge over ignorance, a victory of good over evil, and a victory of hope over despair. Jains celebrate Diwali to mark the attainment of moksha, or nirvana, by the twenty-fourth and the last Tirthankar, Lord Mahavira about 2600 years ago. Sikhs celebrate Diwali as Bandi Chhor Divas, on which day, sixth Sikh guru, Guru Harbind Singh, along with a large number of Sikh religious leaders and Hindu Kings were released from prison by the Mugal emperor Jehangir about 500 years ago.

Diwali is not only celebrated as the single most auspicious day, but it is also celebrated as a long holiday period, called the Diwali season, with almost every day carrying a different significance. Dushera, the day of killing of the ten-headed Ravana, is celebrated twenty days before Diwali. Towering effigies of Ravana are burnt very joyfully. Dushera is also called Vijaya Dashmi, meaning tenth day of the Ashvin month on which the victory of Lord Ram happened. Vijayadashmi is also celebrated in celebration of victory of the goddess Durga over the demon Mahishasur.

Another important day for most people, particularly for businesspeople, is Dhan Teras, the 13th day of the declining moon cycle, two days before Diwali, when the wealth goddess Laxmi is prayed to. Whole families participate in the Laxmi puja, which is the prayer to Laxmi, and chopadapujan, which is the prayer for coins, bank account books, and/or checks. A period of about a month around Diwali is

MY BIRTHDAY AND DIWALI

filled with days which have different significance. The day of return of Lord Rama to his kingdom in Ayodhya after his victory against Ravana is celebrated on the day of Diwali.

I am not expected to have any recollection of my birth, either before or immediately after. But I can still imagine the scene of my birth. The reason I can do so is because of the many movies I have seen in which a similar situation is shown. My birth was in a village, and it was handled by a midwife. So, it can be concluded that delivery must have been painful. Delivery was risky since no medical help would be available if things went wrong. Babuji was not even present during my birth because of his job. But, just like in movies, when the birth was announced, everyone was so happy. Everyone is always happy with a birth, especially if it is a boy. So, I can imagine the joy I must have brought to everyone. I can also imagine all the pain during birth and recovery and the follow up care that was required for Baiji. I can understand why she loved me so much.

I was chubby by Indian standards. I had a full head of dark black curly hair. I was brown, generously referred to as a wheat complexion. My parents and the rest of the family were glad that I was not too dark. The news of my birth was sent to all the relatives by post card, telegraphs, and personal visits by family members. Sweets were distributed to friends and relatives on these visits. If someone came to visit for the first time after my birth, sweets were given to them for almost a year.

There were some friendly disputes among the relatives who visited about whether I looked more like Baiji or Babuji. People close to Baiji always thought I looked more like her, whereas people close to Babuji thought that I looked more like him. I don't know how I compared to Baiji or Babuji as a child, or as a young adult, but I do know at my current age I am an exact copy of Babuji, and I am proud of it. My parents played a greater role than anyone else in my life.

CHAPTER 5

My Parents

ACCORDING TO HINDU religion as well as most religions in the world, God made the world. Therefore, He occupies the highest position. However, there is lot of competition to fill the next highest position. Parents certainly qualify in most circumstances.

Up until the last two hundred years, change of pace was slow. Under such slow change, anyone older than you is likely to know more than you might. Certainly, parents, teachers, and your Guru are older and therefore wiser because of greater experience. More importantly, these three love you and are willing to make lot of sacrifices for you.

For these reasons, and for others, traditionally in India children are taught to respect their parents and obey them, even if children think that they are wrong. I loved and respected my parents, even though my mother had very limited schooling and my dad did not finish high school. I also appreciated the great sacrifices they made to raise nine of us brothers and sisters so well. All our needs were always taken care of to the best of their ability. We all received a good education and a good upbringing. That debt can never be repaid.

Let us start with my mom, Baiji, Subhadra Chhajed. She was born in or around 1918 in Madhya Pradesh, India. I cannot be completely sure of the year because of poor records and the inaccuracy of recollection by me and others after more than a century from her birth. We celebrated her birthday on May 25th each year.

She passed away in 2010 at age 92. She was in good health until the last 2-3 years of her life when she repeatedly had infections and fever. She was a devout Jain. She believed in reincarnation. Based on her circumstances, she decided that it was time for her to be reborn into her next life. So, she refused to eat for a few days, observing a controversial tradition of Jainism, called Sallekhana or Santhara; she passed away after only one day's fast. Santhara, a fast unto death, done for religious reasons under religious guidance, is considered one of the best ways of death by Jains. During this period, you ask for forgiveness from everyone who you may have hurt in speech, thought, or action throughout your life, and also in all your past lives. You are also mentally detaching yourself from this world, and thereby departing without feeling sorry for yourself and others. She was lucky to have done so.

Both of Baiji's parents died before I was born. Baiji's older brother, Bade Mamaji, became the senior most person in the family. She was very proud of him and would always have stories to tell us as to how much she was loved. Whenever she compared her life after marriage to her life before marriage, as many new brides and many wives do in India and the world over, she always maintained that she lived in luxury and was cared for in every way before marriage; but marriage changed everything for the worse. She always thanked Bade Mamaji for it.

She was a tough wife. She was a tough mother as well. She gave birth to twelve children—4 daughters and 8 sons, all during twenty-two years from 1935 to 1956 while she was between the ages of 17 and 38 years old. My oldest brother died prematurely, and my second oldest sister also died at about six months of age. My oldest surviving brother died before he graduated from college.

GROWING UP IN MUMBAI, INDIA IN 1940S, '50S AND '60S

She raised seven brothers and three sisters to the best of her ability and means. All her children loved her. I had one older brother, Nirmal, nicknamed Babu. I have one older sister, Nirmala, nicknamed Baby, and two younger sisters, Urmila, nicknamed Chhoti, and Sheela, nicknamed Doctor. I have five younger brothers—Virendra, nicknamed Viru, Sushil, Dilip, Prakash and Anil, nicknamed Baba. Baiji raised all of us with a lot of help from my sisters for household chores until they got married and left. As my brothers got married, their wives helped. Boys did not help as much. I helped very little.

She was very popular among her neighbors. She was even popular among the children from our neighbors. They all loved her and would stop by just to see her. The difference in age did not matter. This did not change even as she aged from a young woman to 92 years when she died.

She believed in, and followed the Jain religion in her life, and spent a part of her day doing Pratikraman/seeking forgiveness and reading religious books. She visited a Jain temple regularly and listened to religious discourses whenever she got a chance. She cooked in strict compliance with Jain-approved ingredients and methods. There was no meat or egg allowed. Even root vegetables such as onions were avoided. Vegetables with lots of seeds such as egg plants were avoided. She avoided any thing that she believed was harmful to our health. She studied only up to 4th grade, as was common for girls at that time. She could read, but slowly. She could write, but her hand was not steady. It took her long time to finish a letter.

Despite her limited schooling, she knew how to manage money. Her math was good enough for that. She saved money and kept it safe with friends she trusted. She also taught all her children the value of saving and managing money. Because of this, we kept our needs in check. Even though we were a large family, all our needs and desires were satisfied. Until her death, she retained financial control of the household, and therefore was able to live in the same home for fifty years until her death, whether children liked it or not.

She valued education greatly, even though she herself was not able to study beyond 4th grade. She insisted that we study hard and do well in school. Everyone in our family got as much education as they desired. If she had to beat us to get us to study when we were younger, she did not hesitate.

She was hard working. Baiji did most of the laborious housework herself, with a very limited household help. She cooked all meals and snacks, she washed clothes and utensils, she mopped the floor, she kept everything neat and tidy. The reason was to save money and because she was very particular about how things needed to be done. A lot of household help quit because they knew that they were going to be fired sooner or later for not meeting her perfect standards. Food had to be prepared exactly right in flavor and quantity, floors and utensils had to shine after they were cleaned, white clothes had to be truly white after washing. She had no patience for any shortcuts to save labor in anything. She saved money for her own security and maintained a tight control on money and property so that even when she was old, she could not be ignored.

Our home, though simple, was always very clean and organized. We kept all our belongings, clothes, books, kitchen utensils, supplies, and tools very organized; this took up very little space.

Babuji, Jainendrakumar Naharsingh Kothari, my father, was born in or around 1914 in Sitamau, where I was also born. We celebrated his birthday on October 26th. He passed away in 1992. He was in perfect health until last few weeks before his death when he was treating himself for a cold and flu. He did not realize that it was malaria. When things got serious, he was admitted to hospital and treated there. Despite that, about a week later he died. He was otherwise in good physical and mental shape, with no major health problems. It was a most unfortunate mistake. If the malaria had been diagnosed sooner, he might have lived much longer.

Babuji grew up in Sitamau, and then in Mandsaur, both in Madhyapradesh, India. He studied until the 9th grade, when he left school to take up a job in Bombay. Babuji was raised in a family with modest but comfortable means. Babuji's dad, my grandfather, Bhausaheb, had inherited a big mansion in Sitamau, which was our family home for generations. He worked as chief engineer for the King of Sitamau, and supervised engineering, construction, and maintenance of the Sitamau palace and other properties.

In a very short time after coming to Bombay, Babuji found a good job with Kamala Textile Mills, in Lower Parel, Bombay, and rose to the highest position possible in office with a title of office manager. In 1960, at age 46, he was forced into retirement by a new generation of owners who inherited the business. He tried his hand at business, selling textile wholesale with mixed success. During the period my father was in business, he kept his own accounting ledger, which was very time consuming. I helped him sometimes. But when business did not produce enough income, Babuji and Baiji did not get along.

Babuji found excuses to stay away for long periods by himself, sharing our Sitamau house with two of his younger brothers a few years after he retired and could not make it in business. My oldest uncle, Sampat Lalji, bade Kaka Saheb, was a widower for a long time and had gone to Sitamau after his retirement from another Bombay Textile mill, Bradbury Mills. Another uncle, Sohan Kaka Saheb, a retired schoolteacher, never left Sitamau.

Babuji's command of English, Hindi, and arithmetic was incredible. His handwriting was immaculate. His vocabulary in both Hindi and

English was very extensive. He was religious and believed in the principles of Jainism. He attended Jain lectures at a nearby Derasar/Jain temple. He was not very high on religious rituals and Tapasya. We are Sthanakwasi, and do not worship idols and do minimum religious rituals. He read a lot. He loved to explain whatever he had learned or discovered, sometimes even when the audience had no interest. He was interested in politics, religion, philosophy, proper human behavior, music, poetry, and literature. He recorded his thoughts and poetry he had written in Hindi on cassette tape recorders, as if he was speaking to a large group, and also wrote them in handwritten books; he may have thought of publishing, but never did. Maybe when this book is published, it will satisfy his unfulfilled desire.

Babuji tried to maintain high moral standards for himself and he tried to instill these values in us every chance he got. While I was sometimes annoyed, I look back and thank him for that.

Combined families, with more than one married couple living together, is a long-time tradition in India. It is the most efficient system in providing for a large number of family members who are dependent on agriculture on a limited land. This may have applied to about 75% of Indian population when I was born. In combined families, family businesses survive and prosper for generations, as skills and capital are preserved and passed on for generations. Even in families where breadwinners are mostly employees, it was necessary that many generations live together because of rising real estate costs and low income. Combined families also provide a greater safety net for people in their old age, particularly for widows and widowers. If the family is successful, all members share in the fruits of the success more or less equally, regardless of their contribution. Unfortunately, this allowed many to become lazy and have a good life without earning it. While I was growing up, we did not have any married family members except my parents. However, it was anticipated that my sisters would leave after marriage. But all brothers would continue to live in this apartment with their wives and children after marriage, except those who were very affluent.

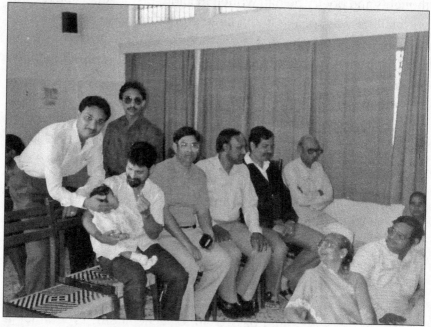

Family Group Picture. 2nd, 3rd and 6th from right: Mother-Baiji, Father-Babuji and Author.

Father-Babuji

Mother-Baiji

CHAPTER 6

Our Home in Sitamau, My Birthplace

AS MENTIONED EARLIER, Baiji traveled from Bombay to Sitamau to give birth to me, in my grandparents' home in Sitamau, Madhya Pradesh, India. Therefore, Sitamau has a special place in my heart.

My grandparents' home was a big two-story castle-like house measuring approximately two hundred feet frontage and 100' deep. It was built by a former King of Sitamau and was built strong like a fort. Exterior walls, facing the main road, did not have any windows. They were twenty-four inches thick made of brick or stone. In lieu of windows, they had open gun ports measuring six inches wide and two feet tall. The ports were set at a 45-degree angle to the wall, such that from the inside you could aim you gun at an intruder who is likely to come from outside the city, but the intruder would not be successful in shooting back at you. This protective opening provided some sunlight and breeze.

Exterior walls not facing the main road were built to more normal standards and had normal windows. The house was built with an open courtyard/aangan in center. Inside walls facing the Aangan had doors and windows. Doors and windows had transoms with diamond patterned openings made from cement that allowed in the breeze and light, even when the doors and windows were closed for privacy. All the light and breeze came from the Aangan area.

All walls were painted white in water-based paint. Some of the kings in Rajasthan chose a particular exterior wall color to brand their cities. Examples were light pink for Jaipur and light blue for Jodhpur, both in Rajasthan, which is an adjoining state with similar culture. Sitamau did not have such aspirations.

The Aangan measured approximately twenty feet wide by forty feet long. The Aangan was the brightest, most airy, and most pleasant place in the whole house. There was very little rain, so the Aangan was in use all day long throughout the year. It never got very cold, but in absence of artificial heat, during one month of winter, January, you shivered a little. The Aangan was the hub of most activities. Indians love company. Indians hate to be alone. This may not be unique to India. Most societies, where income levels are low, and where people need each other for survival, tend to be that way. Everybody walked through this courtyard to go from one part of the house to another. So, if you loved company, you spent as much time in the Aangan as possible.

When you got up in the morning, you chewed on a Neem twig in one corner of the Aangan. You chewed on it for several minutes at one spot and moved it around to cover the whole mouth. Life was leisurely. It took fifteen to thirty minutes to brush all your teeth and gums that way. Those who did so faithfully, had very nice white teeth, fresh breath, good gums, and no dental disease. Neem juice is bitter in taste, kills bacteria, and prevents bad breath.

You ate your breakfast, lunch, and dinner in the Aangan. You sat on the floor cross-legged in a yoga position called sukhasan/happy sitting position. You sat on the concrete floor after spreading a mat, chattai, bed sheet, rug, or carpet. You ate in a round metal plate with a rim called a Thali. Thali was made out of stainless steel, aluminum, brass, or copper. Thali was placed on a two inches tall, small table called a pata. Some people preferred to sit on a pata also. Milk and tea was served in glasses or cups made out of these popular metals. Daal, curry, or vegetables with gravy was served in katoris/bowls made out of these popular metals. Food was always mouth-watering delicious. Always spicy or sweet in abundance. You ate the food with your fingers of your right hand.

Ladies did a number of things in the Aangan. They washed utensils used for cooking and storing food. They washed thali, bowls and glasses used for meals. This was done in a manner that minimized the use of water. The scarcity of water made this absolutely necessary. First, dirty utensils were cleaned with fine sand. Then, they were given a final cleaning and polished totally clean with second batch of sand. After this was done, utensils were shiny. They were then sprinkled and washed with a small handful of water. After that they were wiped clean with cloth and polished with a clean cloth to a brilliant shine. Now they were completely clean and free of any remaining sand.

They also washed clothes in the Aangan. Clothes were washed by first soaking them in soapy water in a steel bucket. Clothes were then beaten with a wooden ram, called a Dhoka, which is shaped like a 2-foot-long baseball bat, against a flat wooden plank. After wringing as much soapy water off as possible, they were rinsed in clean water and water was squeezed out by wringing. This was done several times to complete the cleaning. Then, they were hung on stretched ropes and dried in the courtyard. Sitamau gets very little rain, and the air is dry, so clothes dried fairly quickly. The only problem was that when the clothes were dried this way, they ended up with lot of wrinkles. So if you were fussy, your clothes were also ironed by ladies.

GROWING UP IN MUMBAI, INDIA IN 1940S, '50S AND '60S

All food items were grown locally and purchased fresh. There was no electricity in Sitamau, so obviously no refrigerators. Ladies sifted through grains, beans, and lentils in the Aangan. The grains—wheat, jowar, bajri, corn and gram, after cleaning, were taken to a flour mill for grinding between mechanically-driven or buffalo powered grinding stone wheels. In smaller quantities they were ground manually in a device called a Chakki, in which you manually rotated a circular stone with a handle over another circular stone. Spices, particularly red pepper, were crushed after first being dried in the sun. They were then beating with a wooden ram inside a stone bowl.

Food was cooked in a kitchen, which was barely tall enough to crawl. Gobar/cowdung pancakes were burned as fuel for cooking. The small vent was not sufficient to reduce the high concentration of carbon dioxide, carbon monoxide, and methane in the smoke. People legitimately complain about pollution in the large cities in India. But the pollution in the Sitamau kitchen was hundreds of times worse. I am sure it reduced the lives of women who spent a considerable time in kitchen, including my Dadi. Women used the Aangan all day long, except when older men were going to be present. Men came to eat or meet others in the Aangan. Women always covered their heads and foreheads with saris in the presence of older men, particularly if they were elderly men. This was done to show modesty and respect. Many would cover their entire face by pulling their Sari an additional twelve inches down. This covering is called a Ghunghat. With a full Ghunghat, women could see only a few feet in front of them. To see farther, they pulled the sari in front a few inches, while keeping it low enough to hide their face, and they turned their head down to look through the opening. Most women did not like this, so they sometimes violated the Ghunghat discipline, when they were out of their home. But if an elder reported this violation, and if this violation happened in a public place, the woman would be strongly reprimanded.

OUR HOME IN SITAMAU, MY BIRTHPLACE

The Aangan was also where naps were taken in the afternoon. The Aangan is where you relaxed in the evening, held meetings, and had parties with friends. The Aangan is where men slept on a bed made out of jute ropes. The Aangan is where kids played and studied. That is where I played cards and board games when I visited. There was no place like the Aangan.

As mentioned earlier, there was no electricity in Sitamau at that time. A lalten/kerosene lamp was used at night when you needed a bright light for reading, eating meals, or holding meetings at night. Regardless, your activities were significantly limited at night. Perhaps nature expected you to go to bed early and wake up early. Electricity changed all that. For a night light, you used Diya, a clay cup filled with oil, into which a cotton wicker, baati, was placed and lit.

I can very well imagine how much fun I probably had running around the big house without any clothes on when I was a toddler. I do remember, when I was about five years old, I was eating my lunch and my dadiji was placing hot roti in my thali. She noticed that I was eating with my left hand. She asked me to get up, wash my hands, and come back to continue my lunch. I was annoyed. But I did exactly as she said. Before I could resume, she explained that I cannot eat with my left hand, because the left hand is dedicated to wash yourself after using the toilet. Everything was to be done with the right hand, if one hand is enough. Thus, she succeeded in changing me from being a left-handed person to a right-handed person. I don't think I suffered any grave consequences due to this forced change, but I don't recommend it for anyone else.

GROWING UP IN MUMBAI, INDIA IN 1940S, '50S AND '60S

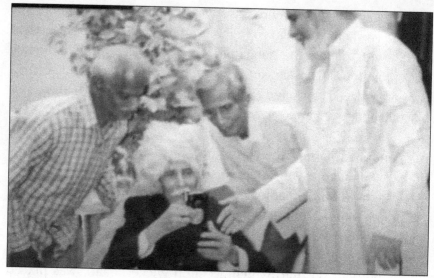

*My three paternal uncles from left:
Sohanji, Sampatji, Dulhesinghji. Brother Dilip on right.*

Our house was directly opposite the entrance to the king's Palace. One of my great grandfathers was adopted by a King of Sitamau. The king gifted our house to this great grandfather. Dadaji, Nahar Singh Kothari, and Dadiji, lived in this sprawling house, which would have been beyond my Dadaji's means to acquire. Babuji and my paternal uncles/kakas were all brought up in this house into their adulthood before they all left to go to Bombay for better opportunities. However, their love for this place never vanished. One by one, Babuji and all Kakas, except the youngest one, spent a good part of their retirement years in this house.

OUR HOME IN SITAMAU, MY BIRTHPLACE

Author's Grandmother Dadiji

The Sitamau State was a small princely state under the British rule/ Raj. In a mega deal, under the leadership of Vallabhbhai Patel, the first Home Minister, almost all princely states were brought into the Indian Union by a combination of coercion, persuasion, and promise of privy purse/retirement funds after the Independence of India in 1947. Sitamau readily joined the Indian Union. Sitamau probably had a population of two thousand when I was growing up. Even so, Sitamau laid claim to a famous personality, our former king and Hindi writer Dr. Raghubir Singh, whom I admired. Sitamau has grown since then, and now even boasts of a college within the village.

My parents had to be brave and confident to leave Sitamau for Bombay. I admire them for that. No one chooses where they are born and from whom they are born. But I consider myself lucky that, perhaps due to my good karma, I was born and raised in a Jain family started by Babuji and Baiji in Sitamau and continued in Bombay. History was going to repeat, when I would grow up and leave Bombay for the United States.

CHAPTER 7

Our Apartment in Mumbai

BABUJI STUDIED UNTIL ninth grade and migrated from Sitamau to Bombay, now renamed Mumbai, at a young age in search of employment. Babuji found a job at Kamala Mills, a textile mill located in the Lower Parel Area of Bombay. This is the industrial area where there were many thriving textile mills. Textile mills were a major industry of Bombay at that time, but the industry has disappeared with changing times. Kamala Mills was closed along with many other textile mills near the turn of this century. Many textile mills were converted into malls. Kamala Mills was converted into a television broadcasting hub. However, all this happened long after Bauji's retirement. While Babuji worked there, Kamala Mills was doing well and employed several hundred employees in production. Babuji progressed steadily and reached the highest level in office. He achieved all this even though he did not complete high school. He taught himself. He learned whatever he needed to learn. He acquired a good knowledge of accounting. He had a good English speaking and writing ability. He negotiated well. He was responsible for managing about fifty employees working in office. He also handled all purchases and sales. He filled out all the checks, which were then submitted to owners for their signatures. He drafted and proofread most of the letters for various purposes and submitted them for owners' signatures.

OUR APARTMENT IN MUMBAI

He found an apartment in a building named Kavarana Rutton Mai Mahal, Kavarana Building for short. The building was named after the developer's mother, Rutton Mai Kavarana. The last word in the name, "Mahal", means palace. To say that the name palace was an exaggeration would be an understatement. It was actually a chawl apartment. Chawl, the anglicized version of the Marathi word Chaal, means sidewalk or walkway. This type of construction was used to provide the most affordable living quarters. The building was not all that different from hotels with an exterior corridor that are very popular along highways in United States. The building was located between two local train stations on the western local railway line. The stations were Dadar and Matunga. The postal office serving this building was named Mahim. Also, there was a large public park called Shivaji Park close by. Being close to the Western Railway and municipal bus routes on the busy Lady Jamshedji road made it very convenient to get around. No matter what outsiders think about living in a big city like Bombay or New York, convenience of public transportation really makes these large cities desirable, even though there is plenty to complain about.

Rent was affordable. Space was big enough for the small family Babuji and Baiji started with. They had no idea then that eventually this small family will grow to a dozen members in the Kothari family, all who would be living in this small, two-room place. And they probably never thought that they would be living there for about twenty-five years before moving elsewhere. By the time we were ready to leave this place, I was 18 years old and had not only finished high school but had also completed two years of science college.

You can understand why a famous building like the White House would have a name. But you may wonder why a building like the Kavarana Building was named. The reason is complicated. In India, street numbering is very chaotic. Homes or businesses do have a numbered street address, but this numbering is not very useful in most places in India, although it is now improving in some of the larger developments in bigger cities. Numbers do not line up in sequential order most of the time. Numbers are rarely displayed. Even if they are displayed, all signs are very small, and they are only readable if you are close and right in front of it. Names of buildings are displayed in larger letters and more readable.

A Chawl Building in Mumbai

When a building is new, owners are proud to display the name, which often refers to family members such as grandparents, spouses, grandchildren, etc. They are also named after one of many gods and goddesses to get their blessings. Many times, it refers to the hopes and aspirations of the owner. Regardless, after a while the sign may disappear, may get covered with dirt and grime, or may become obstructed by a bush or something. If this happens, the owner does not bother to fix it. Since the street address is not displayed and the name is not legible, all you can do is find the place you are looking for by asking a lot of people and, of course, praying to your gods for help. When you do find the place, you are so happy with your good luck and your god's help. Google map, or similar services, may be changing all that.

In any case, if you don't know the place already, and do not know anyone who can tell you how you can get there, all you can do is get as close to the place as you can guess, and then start asking everyone on the street if they know where it is. Be prepared that some of these people who claim to know may not actually know or may not know how to explain it. In any case, since most Indians are pedestrians, they generally think in term of short distances. Also, since many of them spend most of their life in a small geographical area, their descriptions of places are not as complete as you might need them. All they will do is point you in the right direction for about a block or so. After that, you stop and ask again. If a turn is required, the person trying to help you will explain and make gestures, which, don't be surprised, may conflict with what they are saying. Usually it is best to ignore the gesture, but sometimes it is the other way. So, with only the address, you have a chance that you will find the place, and when you do you will feel so lucky. But if luck is not with you, you may come back home feeling stupid after wasting a couple of hours. But don't feel so bad. You are not the only one. It has happened to me many times!

When we moved in, the building was in good shape and good looking. However, our building, just like almost all buildings in India, deteriorated fast in appearance and maintenance. By and large Indians living in cities and towns reside in apartments. Single family homes are just not affordable, except for the very rich. When I was growing up, they only rented apartments. However, most new construction now are ownership flats or condominium apartments. Regardless of whether a building is rented or owned, exterior appearance of the building and maintenance of the common areas is almost always poor.

GROWING UP IN MUMBAI, INDIA IN 1940S, '50S AND '60S

Let us first look at the rental situation, which is what our situation was. There are many reasons for the poor upkeep. The biggest reason is the attitude of people. Indians tend to take care of everything inside their own homes very well. But outside their own home, they generally don't care. To make matters worse, you have the rent control laws, which are intended to protect tenants. Due to rent control laws, rents cannot be raised very much, and so tenants pay very low rent to begin with. Then, over a period of time, the allowed rent increase does not keep up with inflation. So, landlords make very little money from rent after a few years. Tenants don't move because if they did, their new rent would likely to be significantly higher. So, they put up with poor maintenance. Landlords on the other hand prefer that old tenants move out. I will explain why in more detail later. Homeowner's associations try to take up this maintenance work, but not every tenant contributes. Therefore, homeowner's associations are also very ineffective. Thus, like most buildings in Bombay, and all over India, this building was poorly maintained and never painted. The building leaked. The association did half-hearted repair jobs. Some of the tenants who lived with us when I was growing up are still there almost 60 years later.

In India, the free market in the true sense flourishes due to a lack of strong law enforcement. So, one side effect of the rent control is unusual—a one-time black money non-refundable down payment called a Pugri, which is required from a new tenant. This money should really belong to the landlord; but, it is also shared with old tenant to facilitate his moving out. If the landlord has his way, he would love a lot of churn of tenants.

Almost all residential construction in most cities and towns is poured in place with reinforced concrete. They consist of slabs supported on frames, beams, and columns that are supported on footings. This frame is filled in with partition brick walls that are non-loading bearings. Additional dividing non-load bearing brick walls are located per apartment plans, as required. These walls and all exposed surfaces are finished with a cement plaster and painted with bright colors. As a result of a high salt content in the Bombay air, strong and continuous rains (eighty inches in four months), high humidity, strong winds, and poor maintenance, most of the buildings look terrible and leak during four months of monsoon. You see streaks of gray or black mildew and cracked concrete plaster finish everywhere in Bombay and the rest of India. If you can imagine an old sidewalk dark with mildew, you might be able to visualize a typical exterior wall of almost any residential building in India.

The building plan was a rectangle with a rectangular hole inside it. Two buildings—one fronting the main road, Lady Jamshedji Road, and the second one of same size right behind it were constructed and separated by twenty feet of space. The space at each end was closed with an area for common baths, common water taps, common toilets, and a connecting corridor at each floor level. This left a rectangular hole or courtyard between the buildings. The front building had five floors, and the rear one had four floors. The ground floor in the building with the frontage on Lady Jamshedji Road had shops. Both buildings were topped with a flat terrace. A concrete water tank was built in each corner of the terrace to serve each building. Each floor had five identical apartments on each side of stairwell, which was located in the middle of the building, for a total of 80 apartments—40 in each building, and 10 shops/apartment in front building. The stairwells ran in the middle of the building and connected the ground floor to all floors and terraces. Stairs were concrete steps and were arranged in a single flight from floor to floor, and they lined up one above the other. After walking up one flight, you made a U-turn at each floor and walked back to the start of next flight up to continue to the next floor. The landing measured 12′ x 12′ at each end of the stairs. Once you reached your floor, you crossed the landing and made either a left or right turn to go to your apartment. The landing was big enough to serve as a meeting place for kids. Our apartment was on the top floor of the front building.

GROWING UP IN MUMBAI, INDIA IN 1940S, '50S AND '60S

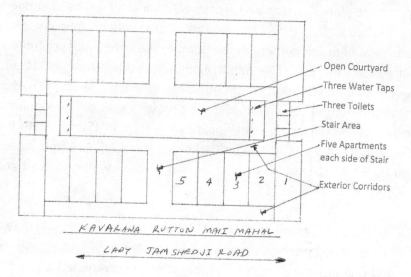

Kavarana Rutton Mai Mahal Plan. 5 Apartments 10'x22' Each Side of Stair Area.

Each apartment measured 10' x 22' and consisted of a living room approximately 10' wide and 12' deep. The living room also doubled as bedroom at night. Behind the living room was a kitchen room which measured 10' x 10', which also doubled as a sit-on-the floor dining room. There was no bathroom. Instead, there was a 3' x 3' wash area enclosed by a curb, called a "mori", which was part of the kitchen. A mori had a tap and a drain. The mori served as a sink in which dishes, pots, and pans were washed. It also served as the laundry tub in which clothes were washed. Since there was no bathroom, the Mori also served as bathroom. Women and girls took baths there. Boys and men generally took a bath outside in the common area between the buildings. A steel or aluminum bucket was filled with water. A tumbler/lota was used to pour water over your body. During cold days, hot water, which was warmed over a stove, was sometimes added for adults. All floor surfaces were covered with granite type tiles. All interior walls in the apartment were painted with glossy oil paint, in any color of the rainbow, in light to medium shades.

OUR APARTMENT IN MUMBAI

Both the living room and kitchen opened up to exterior common corridors, called chawls. There was a 3' high guard wall on the exterior edge of the chawl. The corridor was 5' wide and was used during the day for access to your apartment. It was also used as an extension of the living room, sometimes by placement of some furniture like benches or chairs, and sometimes by use of carpets or floor mats/chattai. It was used at night as an outdoor sleeping area mostly by boys and men. Doors were generally kept open during the day. Windows were generally kept open day and night, except during rain. A mattress, thin and small enough so it could be folded for storage, was spread on the floor, and served as the portable bed used by everyone—inside and outside the apartment.

With so many people needing sleeping space, every square foot of the floor space was used up at night. Mattresses were often shared by family members, friends, or relatives of the same sex, both inside and outside the house. This arrangement worked well when there was no rain. When it started raining, you picked up and moved indoors and did the best you could. During the day, the mattresses were folded, stacked, and covered with a bed sheet in one corner. Tenants of the apartments at the end of the building often enclosed the corridor in front of their apartment and made exclusive claim to this space, even though this was not permitted by code.

There were two water taps on each floor at each end of the building in the Kavarana building, from which rationed municipal water was obtained several hours a day. This water was received at the ground floor from municipal water lines, and it was pumped to all the floors for a short period each day. It was also pumped to two water tanks on top of each building for use during remainder of the period. Most men and older boys tried to take bath at these community taps while the water was running either from direct pumping or from storage tanks. Water coming straight out of the taps during the short period the pump was running was the best available water and was used for drinking, cooking, and of course for making tea.

Hot Indian tea with spices and ginger was consumed all day long in 4 oz servings in a small cup. If you were in a hurry, you poured the tea into a saucer and sipped it with great gusto and a loud sipping sound. Water for immediate use in drinking was stored in clay pots called Matkas, which could hold a couple of gallons of water. Water from matka was really cool and delicious due to evaporation and the carbon dioxide absorbed from the air.

Matka used for storing cool drinking water.

At times, when water was not supplied by Bombay Municipality, water was transported in steel or plastic buckets and stored in steel drums in the rear corridor. This was the water used for bathing and washing.

OUR APARTMENT IN MUMBAI

Under this situation, no one complained much about the quality of water, which was poor. Water for Bombay came from several lakes and was transported after treatment consisting of mainly filtration and chlorination. However, it was carried in exposed steel pipes supported on concrete saddles on the ground. While risk of contamination under this condition was high, Bombay water was generally clean; at least it was visibly clear. Bombay water tasted good to me. It did not have an unpleasant after-taste or odor. You cannot say the same about water found elsewhere in India.

We had electricity in our apartment. We had one fluorescent/tube light, and one light fixture each in the living room and kitchen to begin with. Additional lights and a ceiling fan were added in living room later as more electricity was available.

With only a 3' x 3' mori/wash area among a large number of members in a family, and with tight schedule of schools, colleges, and jobs, it was necessary that brushing be done somewhere else. So, we brushed in the exterior corridor. More sensible members did so in the rear corridor, away from the main road. We tried to go inside to rinse in the wash area inside. However, all the spitting while brushing was done in the rear courtyard, or by some who were not so sensible, in the street. Occasionally, a pedestrian would get a splash of someone's spit, and he would yell at the uncivilized behavior, wipe off his face and go on. The person brushing still continued shamelessly. Indian people are so resilient.

In our apartment, there were one or two lockable steel cabinets with lockable vaults inside the cabinet. Money, jewelry, bank accounts, and other valuables were stored in this vault. The rest of the cabinet was used for valuable clothes, etc. There was a mirror on the outside of the cabinet used for shaving, combing, and dressing up. In the kitchen, open steel shelves were used for kitchen utensils, and were also used to store grains, beans, spices, etc. Large glass bottles or stainless-steel canisters were used for storage. It seems amazing how we could get by with so little and yet feel no scarcity or unhappiness. We felt lucky because we had lot more than others.

CHAPTER 8

Mumbai

Bombay Stock Exchange Building in Financial Center of India, Mumbai

MUMBAI, THE FINANCIAL capital of India and second largest city in India, is made up of number of islands connected through bridges and reclaimed land. In terms of wealth, it can be compared to Manhattan, New York. The Hindi movie industry known as Bollywood is located here.

GROWING UP IN MUMBAI, INDIA IN 1940S, '50S AND '60S

Bombay was under the rule of Konkani fisherman communities known as Kolis and Agris, from about 150 A.D. to 1350 A.D. Bombay was ruled by various rulers, including Muslims, from 1350 to 1530. The Portuguese ruled from 1530 to 1660. Bombay was given to the British rule as a dowry in marriage of a Portuguese princess to a British King in 1660. After that, Bombay remained under British rule until India's independence in 1947.

The city consists of several islands that are connected through small span girder type bridges and an extensive amount of filling/reclamation. Bombay was renamed Mumbai by the Regional Party, Shiv Sena, in 1995. Old Bombay refers to the part of Bombay consisting of the area from Colaba at the south tip to Mahim and Sion in the North. The city has grown on other islands and on the mainland during the 20th and 21st centuries. This expanded metropolitan area is known as Brihad Mumbai, or Greater Bombay, which includes Navi Mumbai or New Bombay.

Bombay is on the west coast of India. Approximately 90 inches of rain falls, most of it generally during the months of June through September. During this period, it rains two days out of three. However, for the rest of the year there is very little rain, maybe a little sprinkle once a month, totaling may be few inches.

The first drops of rain fall in June. This first rain is so joyful for everyone. Getting wet in the rain is not a nuisance. It brings great joy.

More than half the population of India, though the number is shrinking with commercial progress of India, depends on farming for their livelihood. Nearly two thirds of the farming is done without irrigation. So, success and livelihood of farmers depends on a lot of rain. So does the price of food and financial comfort for the rest of the population. In addition, without air conditioning, which is not common for homes in India because of lack of power capacity and equipment and operating costs, almost everyone is miserable under the extreme heat during April and May each year. Temperatures range from 90F to 120F toward the end of May. Every Indian is anxiously waiting for the rain each year at that time.

Many Bollywood (this refers to the "Hollywood" inspired, and "Hollywood" imitating film industry located in Bombay) movies have scenes in which the hero and heroine get wet in the rain. They are ecstatic. They sing and dance in the rain. They fall in love, if they were not already in love. There is always thunder and lightning. The heroine, afraid of lightening, runs and hugs the hero for comfort. The audience applauds the hero's good luck. They also applaud because they are happy to see her through her wet sari or Punjabi dress.

Notwithstanding the romanticizing of rain, it also brings a lot of misery to Bombay. Bombay has an antiquated storm water drainage system. It is insufficient and is not maintained properly. In some parts, the storm drainage system also carries sewage, by design or by abuse. Use of detention ponds to control flooding is almost non-existent. Therefore, flooding of streets is common. At times of heavy rain, it is not unusual that you would walk or drive in knee-deep water at many places. This water is highly contaminated due to dirt, grime, gasoline, leaves, etc. In some cases, the grates covering the storm water are stolen and not promptly replaced. Every year deaths occur because a person or a child gets sucked into the drainage pipe.

The city came up with unconventional solution to flooding. For example, at a particularly busy tunnel leading into Bombay's Chhattrapati Shivaji International Airport, a temporary steel ramp bridge has been built for the entire rainy season so cars could navigate the flooded tunnel.

Rain in Bombay often comes in very strong torrents, and you can get fully soaked in a few seconds while you are struggling to open your umbrella. Umbrellas are not very protective due to the high wind. Frequently, high winds will bend the umbrella inside out and damage it when you need it the most.

People accept all this without loud or angry protest. People in India are very patient, resilient, and take everything in stride, regardless of their religion. The Hindu, Jain, Buddhist, and Sikh religions prepare you to accept all adversities. You are taught that all adversities are predestined, they are caused by your past deeds and therefore beyond your control. Muslims, Christians, and Jews also accept adversities as God's will.

Eventually, all storm water and a significant amount of untreated sewage flows into the ocean surrounding the southern half of India. Contamination can be spotted for hundreds of miles from the Indian coast. India consists of a land mass that crashed into Asia land mass which had China at its southern border. Over last several million years, this India land mass continues to creep forward at the rate of 1-2 inches per year. This has made the border region in the northern states of India more prone to high-intensity earthquakes. Bombay is not in this high-risk region, but still it has suffered earthquake damage. Presence of water tanks at the top of buildings, the weight of brick partitions and exterior walls, and improperly-designed and constructed concrete frames make typical construction more prone to risk.

Areas near the coast are always windy because the ocean is much smoother than land, and ocean water temperature is always different than land temperature. In Bombay, which fluctuates between hot and humid to very hot and very humid, wind is welcome most of the time. It is rarely cold. Except for a few days in a year, you don't even need a sweater.

CHAPTER 9

Life in a Mumbai Chawl

IN THE KAVARANA Building where I grew up and lived until I finished two years of science college, each apartment was occupied by two parents and three to nine children. It was not unusual for two or three relatives—grandparents, uncles, aunts, and cousins, etc.—to live in this tiny space. Even though looking back it seemed awfully crowded, when I was growing up it felt perfectly normal.

I lived in that apartment with six brothers, three sisters, two parents, plus a steady stream of out of town relatives who visited Bombay for vacation, medical needs, or a family get-together. It's hard to believe, but I think this period was the best time of my life.

There was lot of love, closeness, and friendship in our building. Older kids played and entertained younger kids. Older kids babysat younger kids all the time, and always without compensation. They did not even need recognition or appreciation. Older kids walked younger kids to school. Kids helped each other in education. They played together. They enjoyed music, radio, and movies together. They shared their toys, books, food, clothes, money, sleeping space, artistic talents—everything they had.

With kids from eighty families living in two connected buildings, there were plenty of young boys and girls to choose from, and lots of company as I grew up. Feeling isolated or getting bored did not happen very often.

Dadaji and Dadiji spoke Marwari at home in Sitamau. Marwari is the language of people who originated from a region within mainly two current states, Rajasthan and Madhya Pradesh. Babuji and Baiji both grew up speaking Marwari at home and Hindi in school. But Babuji preferred speaking Hindi over Marwari. I and four of my oldest brothers and sisters went to Hindi medium school. Because of this, Hindi became the language of our household. Baiji spoke Marwari with Marwari-speaking women. But everyone else spoke Hindi. My five younger brothers and sisters went to English medium Catholic school. Even so, both at home and in school, they spoke mostly Hindi. Babuji believed that Hindi was a better choice because it was also being promoted as our national language and was the language of the educated class. I never got a chance to learn Marwari as a child, which I regret. My weak attempts to learn this language after I had grown up were not successful. However, playing with kids from chawl, I easily learned Gujarati. Because of my last name Kothari, and a fairly good command of Gujarati, I am often mistaken for a Gujarati. I also learned some Marathi, which I improved with formal classes in high school.

Life in chawls is very different. It is full of extremes. Babuji yearned to move out of the chawl and into a more spacious, more private, more decent living space as soon as he could. Baiji, on the other hand, had mixed feelings about moving out, even when she could, because of love, closeness, and friendship.

Baiji was right. But there are plenty of things that are undesirable about chawls. Crowded conditions, financially struggling neighbors, lack of amenities, lack of privacy, and unsanitary conditions are just a few undesirable things that come to mind. Because of these conditions, there was also a lot of pain, anger, and hatred. There were a lot of verbal and physical fighting. The fighting, however, rarely resulted in serious physical injuries requiring medical treatment, and it never ended in death. But the number of such incidences was large.

One reason for such low physical injury was that the British government had banned ownership of guns for British protection, and Indians had accepted this restriction. Independent India inherited those laws and kept them in place; I think, wisely. Tolerance for differences taught in all Indian religions has also contributed to a lower level of serious violence, despite harsh living conditions. Sometimes the fights were over small amounts of money; but they were mostly over self-esteem. Most people under these circumstances had a low self-esteem, and they would pick a fight just to prove that they amounted to something. Baiji, with her short temper, contributed to some of these fights when she came to our rescue, even though we never asked for her help.

I don't think I initiated any fights. But I did try to defend myself. The most dramatic example of this happened when I was about twelve. We lived on the fourth floor. Directly below us lived a family with a mother who had a temper ten times as bad as Baiji. Because this mother always came to her children's defense, the children were emboldened, and were constantly getting into fights. Shaan, who was slightly older than me, was really bad.

One day I was standing outside in the corridor looking at the road below. Shaan came up to me and accused me of throwing water down onto him while he was standing outside his apartment. I told him that I had not done any such thing. But he refused to accept my denial. He challenged me to go to the stairwell area where we could settle the allegation with a fight. I accepted his challenge.

When we were about to start the fight, I realized that he wore glasses. At that time, I did not wear glasses. Even though we were about to fight, being a gentleman, I gently removed his glasses from his face, set them down in a safe place on the side, came back and then punched him in his face. One punch was all I needed to know that I was going to win. We exchanged a few more blows. And then, Shaan stopped, put on his glasses, and left, promising to me to get even with me later. I thought that was an empty threat, and I forgot all about it. But I should not have.

One day as I was going to my municipal primary school, I was stopped by two adults whom I did not know. They forced me to follow them away from the road a short distance. We were only a few yards from the sidewalk where a lot of pedestrians were walking. But this distance was sufficient for them to know that no one would intervene. One adult pulled my hands behind my back. The other guy started punching me in my stomach. After about five or six punches, they stopped and left. They never identified themselves, and never explained to me why they beat me up. But I knew that Shaan had hired them and had carried out his threat. This experience scared me. After that event, I avoided that road for a while. Even when I resumed walking by that spot, I got shivers every time I passed it for a long time. Eventually I recovered and promised myself to never live in fear.

CHAPTER 10

Primary School Days

ON AUGUST 15, 1947, as I was beginning primary school, India received its independence. I have a very clear recollection of curfews imposed in the aftermath of the Independence due to partition-related violence between the Muslims and Hindus. If you stepped out onto the corridor or balcony, let alone walk onto a street, you could be shot dead. Independent India's birth was painful. I was too young to understand that more than a million died and millions more were uprooted from their homes and became refugees in their own country, on both sides of borders. The pain of this miserable event will not be forgotten for hundreds of years by people who suffered its consequences.

British India was divided into four major nations in 1947 and 1948. Pakistan was formed first on August 14, 1947 and consisted of two parts: East Pakistan, which became Bangladesh after its independence from Pakistan in 1971, and West Pakistan, which was referred to simply as Pakistan after 1971. A new nation of India was formed the next day on August 15, 1947. Two other parts of British India: Burma, now known as Myanmar, and Ceylon, now known as Sri Lanka, got their independence in 1948.

On registration day for primary school, Babuji was at work and could not take time off to get me admitted into the first grade at the municipal primary school which was teaching grades 1 through 7. Baiji was not literate enough to fill out registration forms. So, my paternal uncle, bade Kaka Saheb, took me to get me admitted. Montessori, kindergarten, or pre-kindergarten were not very common in those days. In order to get me started in school, my Kaka Saheb lied about my birth date to meet the official deadline for admission. This false date was conveniently chosen as July 1st, the official cut-off date. Thus, I started school at four months younger than the youngest kid in my class, and almost a year below average age in my class. Despite this fact, I did well in my education. Seeing this, my parents and my teachers requested the school to allow me to skip a grade by taking a promotions examination six months into my second grade.

I still remember taking the examination. I can picture myself sitting on a wooden bench without back support which could seat up to three students in a regular class, and two for this exam. There was no writing surface. All writing was done on a small slate with chalk. Since this was a special exam, there were more teachers supervising than students taking the exam. I passed, so I joined third grade halfway through the academic year. This made me sixteen months younger than the youngest kid in my class, and about two years younger than the average age in my class. After that, I always stayed two years younger than average throughout my education in India until I graduated with a Bachelor of Science in Civil Engineering.

I do not recommend this kind of promotion. I not only missed out on two years of freedom as a child, but being younger by two years than the rest of the class subjected me to much harassment by other kids. This was particularly bad when the teacher stepped out, showed up late for the class, or was absent, and a fellow student monitor was temporarily in charge of the class. It was hard for monitors to stop kids from harassing me. Sometimes, even teachers could not stop the kids from harassing me. Oldest kids were the meanest. They, along with the rest of the class, loved to make me cry. When I cried, tears would flow, and my nose would run. When that happened, the kids took a real delight in my pain and my pathetic looks. So, they would make fun of me even more. This only made things worse. But somehow, I survived. I had to.

The medium of education was Hindi. Tuition was subsidized by the government and was very low. Even so, for the needy, there was a need-based government scholarship. My schoolteacher handed out scholarship forms for us to take home. When I gave this form to Babuji, he was very upset. He scolded me for bringing the forms home. He said, "How could you even think about depriving another really needy child of this scholarship." Babuji was always very high-principled. If he got a chance to give us some good values, he did not miss the opportunity. I am eternally grateful for that.

In the fourth grade, we had better seating. We had wooden desks which were shared by just two kids. The desks had a sloping writing surface and a semicircular groove at the bottom to store and catch pencils rolling down on desk. There was also a level surface at top with a smaller groove to hold pencils. This flat surface also had a small ink holder. The ink holder was refilled by the teacher or class monitor with an ink that was dark purple and was water resistant. If you spilled it on your clothes, which happened a lot, intentionally or accidentally, it was not going to come off. Generally, boys sat with boys, and girls sat with girls. Because I was the youngest, most innocent, and least likely to complain, I was made to sit with a girl who could not be paired with another girl.

There was an annual competition of drama among municipal schools of our region. Each school presented a skit or a song. My older sister, Nirmala, pet name Baby, who was three years older than me had a big part in the play. I had just been promoted to third grade then. I was the youngest in my class and I was her brother. Therefore, I got picked. I just had to come on the stage a couple of times and speak one or two sentences of dialogue. Competition was held in Rivoli Talkies. I remember standing on the stage and looking at several hundred people in the audience. I don't think I was scared. Not bad for my first public appearance.

From Left: Nirmal, Virendra, Author, Nirmala

Clockwise from Top Left: Nirmal, Nirmala, Virendra, Author

GROWING UP IN MUMBAI, INDIA IN 1940S, '50S AND '60S

I had to endure bullying all through the primary school and even partly into high school. Adversity makes you stronger. Perhaps that was the case with me. I studied harder, I was more focused on my studies, and because of that, I did very well in my life. I was not at all afraid when I left India to go to the United States for further studies. I achieved more success professionally, and in business at earlier age. I had the confidence to get married at age 22 and had three children by age 32. I can thank my parents for giving me two additional years of adult life.

I studied a total of five years at the municipal school and left after finishing 6th grade. Except for an occasional hit on my hand by the teacher using the thin edge of a ruler, I did not get into too much trouble. Once or twice, I was made to stand on the bench to teach me what not to do in class. I was never made to sit in rooster position, which is most painful and was reserved for more serious and repeat offenders. In rooster position, after you squat, you bring both of your hands through your legs and then reach and hold your ears. First, it is really hard to reach your ears. Then it is hard to maintain your balance. Quickly, your feet hurt, your back hurts, and you want to straighten out. But you can't, because time is not up. If you quit in the middle, your time starts from zero again. I never got punished this way. But I did see some of my friends having this horrible experience.

I was good in studies, but the first rank was always taken by my classmate name Vijay whose dad was the principal of the primary school. I do not know if he did or did not deserve the first place, but I do know that Vijay did not do as well after he left primary school.

Municipal school did not require uniforms or shoes. I did not wear shoes. Most of my friends did not either. So, it seemed normal. We were warned to come in our best clothes on the day of inspection. An inspector assigned by the government would make what is supposed

PRIMARY SCHOOL DAYS

to be a surprise visit and review school performance and recommend teacher promotions and raises.

I did not get a lot of homework. And whatever I did, I was able to finish it very quickly. So, I had a lot of time to play with my friends.

When I was growing up, there was no TV. There was no phone at home. Elders called and received calls on important occasions or in a case of emergency at a rich neighbor's home. Radio was limited to a couple of hours a day. Elders generally made the choice as to what radio program would be kept on, so we did not spend much time listening to the radio either.

The main activity for all this free time was just to be with friends. Once we were together, we just acted on any suggestions and ideas that popped into anybody's head. There was no pre-planning. There were no appointments. We were simply together, and we did whatever we felt like. Occasionally it could get boring. But mostly it was great. We gathered wherever we could be alone, away from our parents. Favorite places were the corridors in our chawl, open areas near the stairwell, the flat terrace on top of the building, or on the sidewalk in the road.

Many times, we got together in each other's homes. If it was lunchtime, dinnertime, or snack time, or if we were in someone's home, then we would be offered whatever they were eating, and we would readily accept the food without being shy. If it was not mealtime, we would be offered water and some snacks. If it was mealtime, we would be offered meals also. The biggest part of our time was spent talking. We talked and exchanged stories. All Indians love to do that.

Sometimes we read each other's books or magazines. I went to Hindi medium school. But my friends went to non-Hindi regional language medium schools. So, I read their Gujarati and Marathi books.

GROWING UP IN MUMBAI, INDIA IN 1940S, '50S AND '60S

We ran all over the place. We ran through our neighbors' homes. We skated barefoot on tiles which became slick when water puddled up. We watched traffic and anything else that was happening on the road below. We tried to learn to spot car models from a distance and competed on who guessed the greatest number of cars correctly. We watched pedestrians walking on the sidewalks on the road below. Some kids pulled pranks on pedestrians. I never did.

As I got older, I listened to the radio more often. I tried to sing Bollywood movie songs. It was difficult to learn the songs without lyrics and a chance to pause and repeat the songs. There were no tape recorders. Pocket radios and record players had been introduced, but they were too expensive. There were no musical instruments. None of us could afford them. We sang what we could remember. We whistled or hummed the rest. Some of us were good in singing. Others were not. But it did not matter.

Outdoor, we played Aankh-Micholi, Gilli-Danda, Daddi-Mar, Langdi, Khokho, Kabbadi/Hututu, Lattoo, marbles, kites, and of course Cricket. Indoors, we played games such as Carrom, Saap-Sidhi, Lotus/Chinese Checkers, Chess, Antakshari, and cards.

We used soda bottle metal caps and cardboard cigarette cases as play money. The American-made Camel brand was valued higher than local brands. Using this currency, many games were invented by us.

We played many other games that I did not list above. The games we played changed as we got older. But the common criterion was that these were all group activities. They did not require any expensive equipment. Cricket was played with old tennis balls. Table tennis was played with equipment provided in school. These games also did not require any uniforms. In short, participation was free, at least for most of us. Since no money was involved, we did not have to ask for our parents' permission.

As long as we did not get into trouble at school for not doing our homework, we could generally spend as much time away from home as we liked. For the most part, parents were not watching us too closely. They did not question us too much about what we were doing in our spare time. They knew our friends and trusted them to watch out for us. We were playing mostly to kill time. We were not trying, and our parents were not pushing us to excel in sports. We were playing because we enjoyed it and we needed to kill time, which we had plenty of. Most of these activities needed enough children so that teams could be formed. With approximately 200-300 children of various ages within our two Kavarana buildings, we had plenty of choices to make suitable groups.

Now a little information on some of these uniquely Indian games. Please remember the rules for most of these games are informal and vary by players to suit their temperament. Therefore, there are many variations beyond what I am describing here. I am only trying to describe a general concept of the game and not all the rules that go with it. Many of these games are played at a professional or high skill level with written approved rules. They are shown in alphabetical order for convenience.

Aankh Micholi/Find a Hiding Person:

Generally played by children, the catcher is blindfolded and waits while remaining players hide within a limited area in an attempt to avoid being caught. Once positioned, they must stay put. They cannot run away. They make various sounds, mostly to help the catcher identify approximately where they are, but also to confuse the catcher. As soon as the catcher catches one of the persons, this person now becomes the new catcher, and the old catcher joins the hiding players. The game continues until most of the kids have had their chance to become catcher. The game stops when time runs out, when parents stop it, or when the kids stop enjoying the game.

Antakshari/Last Letter Game:

This is played by anyone who can remember some songs— boys, girls, and adults. Two teams are formed. A person starts by singing an opening verse and part of an Indian film song. Any person from opposite team responds with another song that must start with last letter/syllable of the song that was just finished. The singing turn goes back and forth. Failure to sing an eligible song within a short-agreed time results in loss of a point or adding a letter from the word "Donkey". When "Donkey" is completely spelled, the game is over. Everyone makes fun of the losing team. A new session is started, if desired.

Daddi-Mar/Hit Stacked Stones:

This game requires a ball made from cloth or an old tennis ball, and stack of 7 to 11 flat stones or brick pieces, which are readily found in the streets. They can vary in size, shape, thickness, and flatness. They can be stacked in any order.

Two teams—Hitter and Stacker—are formed. Members of the Hitter team take turns and throw the ball from about 8 to 10 feet to knock the stack as hard as possible. If a stack is not hit, next hitter takes turn. Once the stack is hit, the Stacker team quickly gathers and tries to stack the stones back into a single stable stack. While they are doing so, the Hitter team tries to catch the ball and hit one of the members of the Stacker team. If a Stacking team member is hit by the ball before the stacking is complete, this Stacker is out. The stack is restored, and the game continues until all stacked team members are out. Team roles are reversed. The team with highest total number of successful stacking wins.

Gilli-Danda/Short-Long Stick:

This is an Indian version of baseball and cricket, played long before these games were invented. The Indian game may have contributed

PRIMARY SCHOOL DAYS

to some of the components of these two games. Boys generally play the game.

The equipment consists of two wooden sticks about one inch in diameter. A short stick called a Gilli is about 5 inches long and is tapered both ways from mid-point. It is laid on the ground on one of the tapers inside a circle on ground. It is hit by second stick called a Danda which is also 1 inch in diameter and about 15 inches long. After hitting, the Danda is dropped to the ground inside the circle. Traditionally, the Gilli and Danda were made by shaving a tree branch with a knife. Now they can be purchased.

When the high end of Gilli is hit with the Danda, the Gilli flips and spins in air. While it is spinning in air, it is hit with the Danda again as hard and as far as possible. The hitter has to run and touch a pre-determined point.

The Fielder team tries to get the hitter out in one of three ways. The Fielder team players try to catch the gilli. If it is caught, the hitter is "catch out". If the gilli is not caught, the fielder closest to the gilli will throw the gilli to hit the danda which had been dropped inside the starting circle. If the Danda is hit, the Hitter is out. Also, If the striker fails to hit the gilli within three attempts, he is out. If he successfully hits and makes it to the predetermined point, the Hitter gets one point. The Hitter plays again and hits the gilli to start a new point. When all the players in the hitting team are out, the roles are reversed. After all the players on both teams are out, the team with most points wins the game.

Alternately, each time the gilli is successfully hit, the distance to the landing spot is measured with the Danda using it as a unit of measurement. After all the players are out, the team with highest score wins.

GROWING UP IN MUMBAI, INDIA IN 1940S, '50S AND '60S

Kabaddi/HuTuTu/Hold Breath and Tag:

Kabaddi is an ancient highly-skilled Indian contact sport requiring strong lungs, strong legs, and good sprinting skills. It involves fast turns, good speed, quick reaction, great strength, and good teamwork. With so many skills involved, I am surprised why this game is not yet included in the Olympics. This game is mentioned in the Mahabharata.

This game is played by strict rules at professional levels among schools, colleges, and businesses at all levels—local, state, and national. We as kids changed rules as needed.

Two teams of seven players on each team play on a court, preferably of dirt or grass, but also on a black top or concrete. Dirt and grass minimize knee scraping and other bruises. But when we did not have access to such surfaces, we played on concrete or a blacktop road.

A line on the ground divides the two sides—offense and defense. The defense team lines up in a straight line parallel to the dividing line near the end of the court. One player from the offense team, called a Raider, enters the defense side while loudly repeating word "Kabaddi" or "HuTuTu". He is not allowed to inhale in between and cannot break the quick rhythm of repeating the specified word as proof that he did not inhale. If he does, he is disqualified immediately. While holding his breath in this manner and loudly repeating the specified word, he tries to touch as many players on the defense team as possible and return to offense side of the court. Defense, in the meantime, tries to avoid getting touched, and if a member is touched, they and others try to prevent the Raider from returning. Any number from the Defense team may join in trying to prevent the Raider from returning. If the Raider loses his breath before he returns, he is out. If he returns safely, then all the people he touched or who physically participated in keeping him from returning are out. As soon as the offense player is out or if he returns safely back, the roles are reversed.

When one team loses all its members, the team is defeated, and the game is over. Professional scoring and "revival" methods are complicated, and they are not described here.

Kho Kho/Go:

The children form two teams— the Chaser and Defender. The Chaser team players squat in one line in a ready-to-take-off position facing alternately in opposite directions with about 3 feet of space between them. Three defenders enter to start the game. One member of the Chaser team starts chasing from one end of the line. The Defenders start at the other end. The chaser runs to try to tag the defenders, who run around the sitting chasers. They cannot cross the line of the squatting chaser team. The chaser gets defenders out if he touches one of them and the game continues with new defenders entering quickly. At any time, the chaser may strategically choose and touch a squatting member of his team from behind and loudly says the word "KHO" and takes the squatter's place. The touched squatter is the new chaser and pursues the defense members. A defender is out if a chaser touches them. The defender is also out if they exit the limit of the area set by rules or enters too late. Once all the defenders are out, the time is noted, and roles are reversed. The team with the shortest time wins.

Kanchki Goli/Marble:

Within a circle drawn on dirt, gravel, black top, or concrete, marbles contributed by the players in equal amounts are placed. By turns, a player tries to hit the marbles within the circle from a few feet by projecting a hitting marble and wins all the marbles knocked out of the circle. For kids who are right-handed, the right-hand thumb must touch the ground, while the marble which is held between the fingers of the left hand is projected by the middle finger of the right hand in a slingshot like motion. For younger kids, just throwing the hitting marble is good enough. The game ends when all the marbles are gone from the circle. The game is repeated as many times as players feel like.

Langdi/One-Legged Hopping:

There are two teams—the Chaser and Defender. The Chaser team sends one chaser who hops on one foot and tries to touch any defender. The defender is out if they are touched or if they cross the boundary with both feet. The chaser is out if he tires or runs out of time; then the roles reverse. The team with the highest number of tags when game is stopped is the winner.

Lattu/Top:

Lattu is a colorful wooden conical piece with a small flat end at the bottom into which a nail is hammered in. The top is finished to a spherical dome shape. It has grooves on the conical portion. It is painted in beautiful bright colors. A string is tightened around these grooves. This string is used to impart great circular velocity to the Lattu which keeps it spinning for a long time. The team with highest score in spinning time wins.

All these Indian sports were easy to learn to play well enough to enjoy it. However, they were difficult if you wanted to master them. I did not have the patience or talent to excel. In fact, I was often the worst player. Teams were formed by captains selecting the best player first. They selected me usually out of pity, or they did not want to risk losing my friendship. In games where scores were kept or where some kind of currency, such as marbles, cigarette boxes, or soda bottle tops were used, I often got close to bankruptcy and had to borrow to stay in game and needed a lot of loan forgiveness to continue to play.

Walking:

One activity that did not require any special skill was walking and talking. I loved it. Most of my friends loved it, too. We walked a lot.

PRIMARY SCHOOL DAYS

We walked up and down five flights of stairs of our buildings; up and down various corridors; back and forth through neighbors' apartments; up and down neighborhood streets; back and forth to train stations; up and down to Shivaji Park and the nearby beach. As we got older, part of the reason to walk was to admire passing girls. We always had plenty of time, and walking was a good way to spend it.

When I started primary school, I used to be escorted by my older brother, Babu, or my older sister, Baby, who were also in same school. As I got older and more comfortable, perhaps by 3rd grade, I started to walk to school and back home on my own. Our school did not require uniforms. Many kids in school would not have been able to afford them. Shoes were also not compulsory. Most of the kids and I did not wear shoes

We lived on the 4th floor. So to go to school, I went down four flight of stairs, exited the building, turned right, went past a bunch of buildings and shops, passed the City Light Movie Theater, turned right, went past the Rivoli Movie Theater, crossed the road, followed a dirt road for a few hundred yards, and arrived at my school. It should have taken about fifteen minutes, but it usually took longer due to distractions.

On my way back at about 1 p.m. it was hot, and I used to be very hungry. I would pass a shop which sold grains, lintels, masalas, and snack items. They were stacked on a table or in jute bags close by. Most of these items were accessible for shop owners as well as customers standing on road. I would steal a pinch of salt, or a couple peanuts as I went by. The grocer would notice it, and he would warn me not to do it, and then he would smile and let me keep what I had taken.

There were many other distractions on the way. For one thing, there were always broken pieces of the sidewalk or stones from construction material. I would play with these broken concrete or stones. Sometimes I kicked a stone or a piece of concrete ahead of me, and kicked it again when I got there, and repeated this until I got bored.

In the summer, tar on the road would get really hot and soft. Sometimes, tar melted so badly that it would stick to my feet or my shoes.

Sometimes, there were road fights going on. Sometimes, someone would be getting beat up by a mob for picking pocketing or teasing a girl.

Sometimes there were road shows going on. A madari/man with a monkey did a show on the street with all kinds of fun items. I would stop and watch. A crowd of 50 to 100 was not unusual. He would normally do this in evening to catch people on their way home. He collected money from this crowd throughout the show. He closed the show when the crowd started thinning and moved to a new location next day.

There were snake charmers who mesmerized the snakes with their special flutes called Beens. Most people were impressed that the flutes were played so well that snakes could be tamed. They did not know that snakes were deaf.

There were people who sold aphrodisiac pills and ointments that improved male performance. The seller "guaranteed" the product. The language that he used to describe the change in men with the use of the product was hilarious and very convincing for many in the audience. Customers paid and went home with high expectations.

There were Yogis who could levitate themselves after lying on the ground on their back.

There were sadhus who sat cross-legged, floating in air, apparently supported only by one stick held by a fully stretched right arm. The Sadhu sat there. He did not say much. But people dropped donations in a steel bowl in front of him.

Babies, born with deformities, were displayed as miracles and donations were collected.

There were card playing dealers who promised you an easy way to get rich, but most actually ripped you off.

There were fortune tellers who used birds to pick out a fortune telling card.

There were educated "scientific" fortune tellers who read your palms, your face or your feet and told you your future after consulting their astrology manuals.

A woman would be going around in circles rotating and moving wildly, with her team members claiming that her behavior proved that a spirit of a goddess had entered her body.

You could enjoy this vast variety of entertainment without paying a penny. Performers did not mind, because they needed a crowd that would draw others who would pay for the entertainment.

CHAPTER 11

High School Days

PRIMARY MUNICIPAL SCHOOL provided education through grade seven. However, after finishing sixth grade, I was transferred to a private school called Marwari Vidyalaya.

This school was located about a 10 minutes' walk from Grant Road station on the western branch of the Bombay local train system. This train ran more or less parallel to the western coast of the island a mile or so for most part of the western coast of the island until it connected to the mainland.

My parents waited until I was old enough to travel on my own on Bombay trains before transferring me to a private school. In municipal schools, classes were crowded. Teacher quality, teacher preparation, and teacher attitudes all needed improvements. Despite these negatives, I benefited from the exposure to a great diversity of classmates. This diversity involved differences of region, religion, language, income level, caste, skin color, and many others. The biggest benefit was that whenever I met a person with a difference, I rarely reacted with fear. I always assumed that he or she was just another person like me.

At Marwari Vidyalaya, I was a new student and I was two years younger than average student in my class. Most of the other kids had made friends over the six years that they had already spent at the school. I had made none. I was lonely. I missed my municipal school friends. I was not secure in my new surroundings. Some of the kids sensed my difficulty. Instead of helping me deal with it, which would have been nice, they started making fun of me. When I could not take it anymore, I cried. That was a mistake. A grown-up kid in seventh grade is not supposed to cry. So for my classmates, seeing me with tears in my eyes was really funny. They laughed at me. Kids were laughing at me and I was crying. Now I was sobbing, and my nose was running. That was even more funny to them. This happened many times. Sometimes, I was saved by the teacher walking into the class. When that happened, the teacher scolded them for their behavior. But similar episodes happened many times, regardless, for the first two years. However, as time went by, I made more friends, I became more confident, and I was finally able to put this crying chapter behind me.

When I passed 8th grade, I was ranked second in my class. I received an award for this at an annual function. I received three books. One was an abridged Hindi edition of Mahatma Gandhi's autobiography *Experiments with Truth*. Second was a fictional novel by our first national writer named Prem Chand. This novel was about the plight of Indian farmers' miserable life of hard work, poverty, perpetual debt, and servitude to the Zamindars and landowners. Unfortunately, their plight in independent India has not improved a whole lot. Third was a book of Hindi poetry. I am amazed at the teacher who selected these books. It was an excellent and a challenging selection. Before this, only books that I read completely were textbooks. I read newspapers and magazines whenever I could. I was flattered that the teacher thought I was capable of reading, understanding, and enjoying these books. Reading these books widened my world, my interests, and increased my curiosity and my lifelong desire for knowledge and reading.

GROWING UP IN MUMBAI, INDIA IN 1940S, '50S AND '60S

I always received special attention from my teachers. I was the youngest and nearly the smartest. I often gave answers that very few could. As a result, I was praised quite a bit. One of the teachers that I impressed taught us Sanskrit.

Sanskrit was a mandatory subject in high school when I was studying. Sanskrit has a revered status for Indians, like Latin is to Christians. Both languages are not in general use today. However, both are necessary if you want to understand the many ancient literature and religious books written in these languages. Purans, Vedas, epic poems, Ramayan, and Mahabharat, and a vast amount of ancient Indian literature revered by Hindus, Jains, and Buddhists have been written in Sanskrit. Sanskrit was a language used, at least partly, by the highly-educated Hindu, Jains, and Buddhists in India from about five thousand years ago until about five hundred years ago.

The Mughal rule after 1500 A.D. and the British rule after 1750 A.D. diminished its use significantly. The Mughal period helped in creating two new languages—Hindi and Urdu—from a combination of Sanskrit, Arabic, and Persian. These two languages are currently the widest used languages in India, Pakistan, and Bangladesh. They serve along with English as the primary languages that bind these nations, despite so many languages in current use in these countries.

Scripts for these two languages are different. Hindi uses a script called Dev Nagari, which derived from Sanskrit, and is written from left to right. Urdu is written in Arabic script from right to left.

Sanskrit is a very structured language with very precise rules governing script, grammar, and pronunciation. Unlike the most widely used English language, there are very few exceptions to the rule of Sanskrit. This made it easy for me to learn Sanskrit. However, there is also one peculiar feature that makes it very difficult to read and understand. The feature is called Sandhi or compounding of words. In English, a comparable word would be a hyphenated word. In Sanskrit, two or more words are combined into one word, both in spelling and pronunciation without any separation marker such as hyphen or space. So, to understand such compound words, you have to be able to mentally break such words into its parts as you try to read, write, or understand the compound word. This is difficult. For example, if you add 5 and 5, you get 10. That is not difficult. But if you have to break up 10 into two components, there are five right answers. It was tough for other kids in class, but it was easy for me.

Scholars probably invented sandhi to save time when speaking or to use less space in writing. But it also may have been done to impress the less educated, and possibly to exclude them from acquiring knowledge that was stored in Sanskrit literature. I believe if they had not made this choice, Sanskrit may have survived today. Once my teacher was so impressed with my answer in Sanskrit class that he gave me a nickname or perhaps a title—Kashi ka Pandit/scholar from Benares. For next three years until I finished high school, this title stuck with me.

Many kids came to school because their parents would not give them any choice. Others came because they wanted to learn and succeed in life. I was one from the second group. For me, more than success, I just wanted to learn.

India is one of the oldest countries. India has also been one of the least mobile countries until end of the twentieth century. Lack of mobility and a high level of poverty has contributed to strong families and strong community ties. These factors may have also contributed to the presence of a lot of small kingdoms when India got its independence in 1947. These factors lead to a great variety related to culture, region, religion, caste, and class. This also led to the development of thousands of languages. These languages blended together to create an even greater number of dialects. When I traveled in a train when

GROWING UP IN MUMBAI, INDIA IN 1940S, '50S AND '60S

I was young, as I looked at the people who were getting on board at each station, I could see significant changes in dress, language, and attitude. The change was gradual, morphing from one to the other right in front of your eyes as you traveled from station to station, which are, on average, about a half an hour, or 15 miles apart.

As transportation and communication improved, by the time India became independent, the number of languages dropped to several hundred and number of dialects dropped to several thousands. The number of languages is continuing to drop, as greater use of Hindi and English is promoted by all governments, and one regional language is promoted within each state and taught in each state. In addition, local, interstate, and international trade and travel has increased. This has further reduced the use of local language in favor of Hindi, English, or the major regional language.

In order to promote national integration and progress, the central government required that all schools must teach Hindi and English when I was in high school. Those studying in English and Hindi also had a requirement to study a regional language. My school offered Gujarati and Marathi, which was spoken by the two largest linguistic groups in Bombay. This was before the state of Bombay was split into two states—Maharashtra and Gujrat, along linguistic lines. I knew Gujarati pretty well because I learned it from my neighbors as I was growing up. But I did not know Marathi that well. So, I chose Marathi. I could get better grades with a lot less effort if I had chosen Gujarati. But I was more interested in learning a new language. These courses improved my ability to read, write, and speak Marathi. My command of Gujarati, Marathi, and English, in addition to my mother tongue of Hindi, helped me a lot in making friends. I was also able to study from English, Gujarati, and Marathi texts, when good Hindi textbooks were not available on any subject.

I lived in the Mahim post office area. The nearest train stations were

HIGH SCHOOL DAYS

Matunga and Dadar, on the western branch of the local trains. Matunga was about 15-minute walk, and Dadar was about 30-minute walk. I took the southbound train, running between Virar on the north end and Church Gate on the south end, to go to school. Trains were referred to as slow train, fast train, and super-fast train. The slow train stopped at every station, the fast train at fewer, and the super-fast train at even fewer. Church Gate was always the last stop on the south end. But on the north end, the last station in the order of increasing distance from Churchgate could be Bandra, Andheri, Borivali, or Virar. The distance covered for these four stations were approximately 15, 20, 25, and 40 miles, respectively.

The slow train that suited me best would start at Bandra and made stops at Mahim, Matunga, Dadar, Lower Parel, Mahalaxmi, Bombay Central, Grant Road, Charni Road, Marine Drive, and ended at Churchgate. The average distance between the stations was one mile on the south end, gradually increasing to two miles at Bandra, and gradually increasing three to five miles between Andheri and Virar. To go to Marwari Vidyalaya, I boarded the train at Matunga and got off at Grant Road, a distance of approximately 10 miles, which took approximately 30 minutes, including all stops at an average train speed of 25 miles per hour. The fast train took about 20 minutes. Add the additional time to walk from home to the train station and from the train station to school, and it would be about an hour for the trip each way, door to door.

Coming home, I took the train from Grant Road going North. But many times, the train would be so full that it was impossible to get in. To solve that problem, I would go in the opposite and wrong way south to Church Gate where everybody would be getting off. That way I would get a window seat and ride the same train in reverse direction heading north and happily pass Grant Road on its way to Matunga. When I did that, it would take longer, about an hour and half, door to door.

GROWING UP IN MUMBAI, INDIA IN 1940S, '50S AND '60S

The tough part in using trains is getting on and off at your station. Sometimes, when I was trying to get onto the train, the train would start moving before I could get on it. If that happens, you grab the vertical handlebar outside the train, and keep running with the train until your feet pick up speed nearly equal to train's speed, and then you jump on to the train. There was a more difficult problem when I was trying to get off the train because the train started moving before I could get off. If you were not experienced, you probably would not be able to get off and you would miss your station. But if you were experienced, like I was, you would start to make move toward the exit a couple of stations ahead. Even with this precaution, sometimes you would make it to the exit door only after the train was already in motion and leaving the station.

I would get off the moving train, hopefully before it picked up too much speed. It is important that you know how to do it. Someone who has not learned the art of getting off a running train is likely to just jump. If you do that you will definitely fall onto the concrete platform. If you are lucky, you could break your fall with your hands. If not, your head could hit the concrete and you could be severely injured. You must get off the train facing in the same direction as the motion of the train. You prepare yourself mentally to run at the speed of train in the same direction as the train. You should land on your foot which is further from the train onto the platform and begin running with the train. Then let go of the train, and gradually bring yourself to a stop. Hopefully, there is no one in your way while you are doing all this.

Marwari Vidyalaya was an all boys' school. When I was in school at that time, it was pretty common to have separate boys' and girls' schools. Girls' schools were generally smaller and fewer in number. The reason was that there were fewer girls in high schools and colleges than boys at that time, and it is so to some extent even now. There were many reasons for this. Many parents did not think that spending money on a girl's educations was as good an investment as a boy's education. So, if they had to choose, then girl would be kept back. Generally, after a boy grows up, he will take care of his family as soon as he starts earning. He will also take care of his parents in their old age. On the other hand, a girl would get married and leave the household and would be freed from any responsibility for her parents, brothers, and sisters.

Parents were also concerned that their daughters would be corrupted by modern education, modern values, and the liberal environment of co-educational school. They were concerned about teasing, safety, or more serious violations by boys. The net effect was that fewer girls went to school, and if they did, they were sent to girls' schools.

An opposite logic applied in the case of boys. Parents wanted to make sure that their sons got the best education and had the highest financial success possible. They thought that the environment of all boys' school would be best to achieve this. Society therefore preferred to keep the two sexes apart at that time.

Marwari Vidyalaya required uniforms. Half sleeve white shirt and white shorts. Unlike English medium schools, a tie or a bow tie was not required. New uniforms and other clothes would be stitched by a tailor who performed the service at our home. We provided the cloth. He took measurements and sewed them with his foot paddle driven Singer sewing machine. The fabric was purchased in large rolls. Leftover small pieces were used to make other clothes and diapers for babies. If clothes that my older brother had outgrown were available, I would inherit them, and they would be passed on to even my next younger brother if they were in good condition.

CHAPTER 12

Trips to Sitamau

WE WOULD OFTEN travel to Madhya Pradesh during the summer vacation, which began in the middle of April and ended in the first week of June. This is the hottest period of the year. Not only that, but Madhya Pradesh is hotter than Bombay. We always visited Ratlam, because that is where Bade Mamaji, Baiji's older brother, lived. For Baiji, who lost her parents at an early age, Bade Mamaji's house was a substitute for a parent's home, which is called maika/mother's home. Her younger brother, who was in another city close by called Jawara, would always visit Ratlam to be with us when we visited. Sometimes we visited Jawara also. Another place we visited was Babuji's home in Sitamau.

When I was in summer vacation after seventh grade while we were in Ratlam, my Dadiji made strong requests for me to visit Sitamau. This was a good opportunity for some adventure for me. I had always wanted to travel on my own. I was comfortable doing so since I commuted by train to school in Bombay. My parents were also confident that I would be fine. In any case, they were going to make appropriate arrangements. So, they agreed to let me travel alone to Sitamau. I was put on the train from Ratlam to Mandsaur. From there I would be taking a state bus to reach Sitamau.

TRIPS TO SITAMAU

After a few hours' travel from Ratlam, I reached Mandsaur. It was about three hours before the bus would arrive and depart from Mandsaur to Sitamau. So, I walked around the bus station, ate some food, looked at merchandise in various shops, and just tried to pass time until my bus arrived. I had a stomachache, and I needed to go. There was no toilet facility at the bus station, which is typical. So I headed for a private spot on a vacant property next to the bus station. There were no trees or bushes to hide behind; just a few weeds. So, I faced away from bus station, squatted, and relieved myself. If I did not see people seeing me, then I could pretend people did not see me. If they did not see my face, they would not know who I was, and that made it ok. I came back to the bus station, and when I checked around as to how long before the bus will come, they told me that bus had come and gone. Evidently the stop was a shorter stop than I thought, or the bus arrived sooner than I expected, or I did not keep checking time with someone with a watch. No matter, the bus I had been waiting for, for about 3 hours, had come and gone!

The next bus, which was the last bus for the day, was in the evening. So, I had no choice. I waited again. I was at the bus station for about six hours more. There was no phone in our home in Sitamau, so I could not inform anyone. About an hour before departure time, I was surprised to see my second oldest uncle, Sohan Kaka Saheb appear. When I did not reach Sitamau, he came to check on me. He was relieved to find me at the Mandsaur Bus Station.

This was 1954. I was 11. India received its independence on August 17, 1947. India was declared a republic under a new constitution of independent India which was adopted on January 26, 1950.

During the two and half years after independence, a number of things were done to transition from British India to an independent India. All kingdoms were asked to surrender their kingdoms in exchange for a privy purse, a pension agreement, a patriotic appeal, or some other consideration. These agreements were negotiated on a case by case basis by a new transitional unelected government of India appointed by the congress party under the leadership of the home minister and deputy Prime Minister Vallabhbhai Patel. The world's tallest statue titled "Statue of Unity", at 597 feet in height, was inaugurated in October 2018 honoring Mr. Patel near the Sardar Sarovar Dam. Mr. Patel's mission was accomplished, and a fully-united India was created. These agreements provided for a lump sum amount in various combinations of real estate, precious metals and stones, jewelry, and cash, along with an annual privy pension income.

The King of Sitamau also gave up his kingdom in this process. Like most kings, kings who were in power were allowed to run the government until the first election. In any case, people still treated them as kings and showed due respect in their presence. Many of them were also given a membership in the leading political party called Congress, and many were also promised a ticket to run for political office. Sitamau King did run for office and became a member of the state legislature and oversaw the local government that ran Sitamau.

The bus arrived in Sitamau. The bus stopped for about a minute to drop me and one more passenger about half a mile from the entrance gate to my village, and then continued further on the highway. We walked on a dirt road and entered the former kingdom of Sitamau, now a village of Sitamau. The castle was immediately to the right of the road. Our home was immediately to the left. The kingdom was gone, but glory of the gate was still there. The gate was guarded only after dark. To make sure gatekeeper did not go to sleep, the gatekeeper was provided a stool with no back support. As part of his job and to let everyone know that he was on the job, he was required to toll a large bell hanging near the gate every hour throughout the night. The number of tolls was equal to the number of the hour. You heard the bell clearly from our home, and probably from most of the village at night. Otherwise, there was absolute silence, since there were no vehicles, no streel lights, and no street traffic. There were fewer than ten cars, I suspect, in the entire village at the time of my visit—about half of them belonged to the king.

Bullock Cart Transportation in Sitamau

The primary mode of transportation was by foot. However, there were bullock carts used for freight or as taxis for longer travel. Prosperous people used single horse carts called Tangas, which could comfortably carry three people. But since there was not enough business for them in the Sitamau, I hardly saw them. And yes, there were some bicycles. Only men used bikes, if they could afford one. Women never rode a bike.

The primary traffic in the early morning was that of young ladies and young girls going to a well about a mile from the gate, and about half a mile on the other side of highway to fetch water. This water was to last all day for drinking and for all other uses, including bath, washing utensils, watering plants, etc.

Wells would go dry at least part of the year. Then there was no water. When that happened, water was purchased from a person called a Chisti who carried water in a big leather bag on his back.

After I settled down a little bit, I decided to venture out one morning on my Kakaji's bike. As I mentioned earlier, this was also the time when women and girls fetched water from the well. I was not very good at riding a bike. My bike ran into a girl who was balancing a clay pot, called a matka, on her head and carrying another pot balanced on her waist. Needless to say, she dropped both matkas, and they shattered immediately. She fell to one side. I fell on the other side. I apologized. She knew I was not local. You always treat your guests with respect. She laughed and left. Neither of us was hurt.

Village well like this one was the source of water for Sitamau.

I have no idea how the news of my misadventure reached my Dadiji, who was amused and kept this story from Dadaji. She was not going to take any chance that Dadaji might scold me or punish me in any way. Knowing Dadaji, I don't think I would have been punished. My dadaji, we called him Bhausaheb, was a very nice person. He was hard of hearing and talked very little and spoke very softly. People talking to him had to shout. My dadaji was a civil engineer who helped with the design and construction of minor repairs and renovations to the castle buildings. He did not have any formal education in engineering. He had learned his art through apprenticeship and self-learning.

My conversations with Dadiji and Bahusaheb were short: they were mostly about meals. But just looking at me made them happy. I did talk to Sohan Kaka Saheb, who stayed in Sitamau, a lot. Sohan Kaka Saheb, or Kakaji, was the second younger brother to my dad. He dropped out of 8th grade after failing a couple of times. I was about to enter 8th grade. I was a good student and I studied in Bombay where teaching standards were much higher. So, in the matter of education, I was probably way ahead of him. Even though, he was much older than me, but we talked like friends. He respected me and I respected him. He was the principal of a government-run primary school in Sitamau that taught kids up to grade four. He was not married. He also wanted to make my trip as pleasurable as possible. So, he made sure that I could do everything that I wanted to do. He also made suggestions and devoted all his spare time so that I could experience the village life of Sitamau.

Kakaji would take me to the farms where many fruits and vegetables grew. My favorites, as almost every Indian child's favorites, were mango, tamarind, jamun, sitaphal, chikoo, peru, cucumber, corn, and many more. Some we got from the farm, and some we purchased from vendors in the street.

Mango: The mango, called Aam in Hindi, is truly an Indian fruit. It originated in South Asia in an area covered by British India which included Pakistan, Bangladesh and Shri Lanka, and surrounding countries including the Philippines. It is now cultivated in many tropical countries. You will find references to this fruit in Vedas, which were written more than 5,000 years ago.

The mango is India's national fruit. India produces the widest variety of mangoes and accounts for about half of all mangoes cultivated in the world. Mango trees live a long time. Some species produce fruits even after 300 years. However, growing mango trees takes patience. They may take fifteen to twenty years before they produce fruit. It is said that a grandfather, who is foresighted and patient, will plant a tree, so that his grandchildren may enjoy the fruits. Such long-term planning has become so rare now.

Mango leaves start out as orange-pink, change to green, and then turn dark green with age. They remain green forever after that, never turning yellow or changing color. The fruit takes three to six months to ripen. The ripe fruit varies in size and color. Generally, they are green, pinkish green or orange-green to start with. When they are fully ripened, they are yellow, orange, red, but may also remain green. They carry a single flat, oblong pit that can be fibrous on the surface, and has a hard shell. Pulp does not separate easily from the fibrous pit. So, you either slice it off with a knife or bite it off with your teeth. Ripe mangoes give off a distinctive sweet smell. Inside the pit you find a single seed.

Every part of mango is used, either consumed as is, or turned into pickles, chutney, jams, or added as part of various recipes. Unripe mango is a particularly favorite of children due to its sour taste. Stealing unripe mangoes from trees and eating them immediately is such a pleasure for children and often even for grown-ups. Kakaji helped me steal unripe mangoes, and both of us enjoyed them. Unripe mangoes are used in many recipes by chopping them into small pieces. Just thinking about it makes my mouth water as I am typing this. Readers who are familiar with this fruit are probably experiencing the same as they read this. My favorite mixes were Daal and Bhel, with really small cut pieces of sour unripe mango. Ripe mangoes are sweet and filling. Diabetics should avoid mango.

Mangoes come in many varieties. Some are easy to slice and cut into pieces. Others are best consumed as juice.

You can eat mangoes in a number of ways. Let us start with the messiest, but most fun way of eating a ripe mango. You gently squeeze all around to soften and liquefy the pulp, Then, you peel open a small hole where the remnant of the stalk is and suck the pulp through the small hole, while gently squeezing the mango. Squeezing too hard may burst the fruit all over your clothes. Even if you are careful, be prepared for mango juice getting all over your face, and you will also get at least a little bit of it on your clothes. But in the end, all that trouble and mess is worth it. The sweet taste of a mango eaten in this manner is heavenly.

A less messy way to enjoy mango is to collect the juice and serve it chilled with a meal. It goes very well with puri, or layered roti which is served on special occasions, or even regular roti, paratha, or rice as part of a special meal. Another way that takes less effort is to bite and squeeze the pulp from a sliced mango, while withdrawing the skin away and out of your mouth. A clean effortless way of eating it is to get someone else to cut the mango pulp into as large of pieces as possible, and then you conveniently eat them with a fork. Another effortless way is to slice a mango and eat it by scooping the pulp with a spoon.

Imli/Tamarind: Tamarind, derived from Urdu/Arabic Tamar Hind, meaning date of India, has a thin brown shell, which you peel off with your nails, and then dissolve strong sweetish sour pulp in your mouth very slowly and suck it for the longest time possible. When you are done, you spit the pit out.

Jamun/Indian Plum: Jamun is a fruit perfectly sized so that you can drop it into your mouth, squeeze the pulp and spit out the pit. If you eat Jamun, you cannot hide it, because your mouth becomes purple.

Sitaphal: Sitaphal, named after goddess Sita, wife of Lord Rama, is full of pits, at least ten or more. After peeling the dark green skin with spherical protrusions, you enjoy the sweet pulp, while constantly spitting out the pits.

Chickoo: Chickoo was the easiest, tastiest, and most convenient to eat. You took about four bites, spit out one pit and you are done. There is no mess eating a chickoo.

Peru/Guava, Kakdi/Cucumber and Barbequed Bhutta/Corn: Indians love strong tastes. Most Indians don't go for subtle flavors or bland food. They like their tea strong with a lot of cream and sugar. They like their sweets really sweet and their salty foods really spicy with lots of red pepper and salt. So, these fruits/vegetables which are tasty just as they are with their subtle sweetness, are often eaten after sprinkling a salt and red pepper mix and rubbing a sliced lemon on them.

Kakaji enjoyed my company and I enjoyed his. The age difference between me and him did not matter. I was enjoying every minute of my trip and I wanted to see and experience as many different things as I could. So, I asked him if I could go with him wherever he went, and he took me everywhere he needed to go, or where he thought he should take me. He took me to our farm, the marketplace; he took me with him when he visited his friends. We would leave in the late morning after lunch and return only because it got dark. There were no streetlights. There was no electricity even in the homes and shops. So, it was pitch dark after sunset. We would go prepared with a kerosene lantern if we knew we were going to be coming back after dark. But sometimes we came late unplanned, and then we just walked in the dark. I started wearing shoes by then, so I was better protected. He did not wear shoes, but he seemed to be fine.

We traveled mostly on foot and mostly during the warmest part of the hot days. We would sit under the shade of a tree when we got tired or hot. It was common to find shady trees along our way, which was different than in Bombay where you saw very few trees. And in Bombay, most of the trees were recently planted and uncared for.

Trees were often used as gathering places because it was cool under its shade. There was often a circular sitting area prepared under the tree canopy. Sometimes this sitting area was covered with concrete. People would squat, sit cross-legged, or would lie and take a nap under the tree. Or, they could sit on the circular wall at the edge of this platform with their feet dangling on the outside of the wall and converse with other people. Life in Sitamau moved at a leisurely pace, so there was plenty of time to just sit, do nothing and enjoy the shade under the tree. I remember sitting under just such a shady tree, with no one around except my Kakaji. The quietness, the green farm, the deep blue sky, the fresh air, and the cool breeze made me feel like I was in heaven.

On days that we stayed out after dark, it was really dark since there were no streetlights. The sky had more stars. Stars were brighter than I had ever seen. There were so many different sizes, brightness, and even colors of stars. I could see how they were helpful for navigation since ancient time. India has a long history of interest and knowledge of astronomy. Even as early as the Indus Valley Civilization of the Bronze Age, the study of astronomy existed. Many advanced concepts of cosmology, including estimates of sizes of heavenly bodies, their positions and distances, are mentioned in Vedas.

A kerosene lantern was used for light for practically every activity at night. You took your lantern with you when you walked outside your home. Within our home, the lantern was used to provide light around which we gathered to talk. We read under the light of the lantern. We ate our dinner in this light. Because of this, everyone went to bed early and woke up early before the daybreak. Sunlight, when available, was fully utilized.

There was no public transportation in Sitamau. You could ride your bike or you walked. My Kakaji just walked. And he could walk. Let me give you an example. One time when he came to Bombay, he made a round trip of almost the full length of the city from Mahim, where we lived, to the South end of city, including Churchgate, the Bombay docks, and the Napean Sea Road, on foot. He must have walked about 40 miles that day.

When I asked why he did not take a bus, he said he was looking at the shops, and he was having too much fun, and time just passed. So, when I went out with him, I had to walk long distances. I used to get tired. But Kakaji tried to find ways to take care of the situation. One time, he arranged for me to sit on a donkey that was carrying some supplies. Do you know anyone who has taken a donkey ride? No one, right?

Another ride that I will not forget was a long ride on a dark black Indian bull called Saand. This Saand had big horns. He was mean-looking. He was big and strong. His skin was unexpectedly very silky. I sat on him, and we covered several miles that way. I treasure the memory of that ride. As I am typing this, I feel like I am back in Sitamau riding the Saand.

While we are on the subject of riding animals, let me tell you about my experience with riding other animals. This was a couple of years later when I was in 9th grade. Our school had arranged a trip to Matheran, which is a hill station near Bombay in Maharashtra. From Bombay, you took a train to Neral. You changed the train to continue

TRIPS TO SITAMAU

to Matheran. We were a group of about 20 kids. Our teacher had arranged for our first horse ride. Anyone who enjoyed it could go for more rides. Each student paid his own expense for this adventure. On the first trip, we were led by two trainers/escorts. One was walking and guiding from the front, and the other was walking in the rear. Both of them were also giving training and making sure that everyone was safe, and nobody was left behind.

Before we started, they gave us some instructions on how to instruct the horse to start, take turns, stop, and other horse-riding tips. After the instructions, we took an hour-long round trip. I felt pretty comfortable at the end of this first trip. So, I took more rides. In each trip, I improved and felt more confident. I enjoyed the entire experience even though I was very sore for a couple of days.

In another trip, I got a chance to ride an elephant. Riding about eight to twelve feet in air is a thrilling experience. Looking down on people from that height and being able to see great distances also made me feel like a king. I understood why a king would choose to ride a nicely decorated elephant in a kingly procession. In that high position, you are "above" everyone; you are looking "down" on your subject and they are looking "up" to you.

On another occasion, I got a chance to ride a camel, which was equally exciting, but less comfortable. When one of these animals gets up from sitting position, you go through a wild rocking motion each time a leg is straightened.

I sometimes got bored during my summer vacations, but not during this vacation in Sitamau. When we were home, many times my Kakaji and I played board games like chess, lotus, carrom, playing cards, parcheesi, snakes and ladders, and many others. This vacation was one of the most memorable for me. After this trip, I knew where I came from and was proud to know that.

CHAPTER 13

Trip to Madhya Pradesh

ENCOURAGED BY MY happy experience during my trip to Sitamau, I now wanted to know more about my extended family in Madhya Pradesh. In particular, I wanted to get to know my numerous cousin brothers and sisters. After my trip to Sitamau, I never missed any opportunity to travel to Madhya Pradesh.

I visited my older Mamaji, who was a like a father to Baiji because he raised her after her loss of both parents at an early age. He lived in a big mansion in Ratlam, Madhya Pradesh. The mansion was made of thick stone walls which kept the inside of the house cool even in hottest summer days. It was almost like you were in a cave that stays cool no matter what weather is like outside. My bade Mamaji was a lawyer. He had a very successful practice. He had started practicing under the British legal system. After the independence, many things changed, but much of the legal system continued. He was a busy man and he was always in meetings. So, I did not get to talk to him much. When I saw him, I greeted him by touching his feet. He would ask me how I was doing. He would ask me if I needed anything. And if I did, he told his servants and other employees to take care of me. Beyond this, I could observe all I wanted. But I did not have long conversations with him. My badi Mamiji died earlier. I don't remember anything about her.

Among his properties, he had a big farm, which was managed by employees, temporary labor, or contractors. I visited the farm a lot. I also read a lot when I vacationed at Bade Mamaji's place.

Bade Mamaji smoked a lot. He smoked Indian Bidi. Bidi is tobacco rolled in dried leaves. As a result of the heavy smoking, he developed throat cancer. He died from it while he was in his late forties. Even though he died sooner than he should have, death in forties and fifties was not that uncommon among Indians due to poor sanitation, poor diet, high stress, lack of education, lack of medical understanding, lack of medical technology among health professionals, and lack of good medical facilities. While there has been great improvement since then, many of these problems still persist in India, particularly in villages. Bade Mamaji's death was a big blow to Baiji. She remembered and talked about him a lot after his death,

My chhote Mamaji was different. Making money was not important to him, even though he definitely needed to: he was raising seven children—five boys, two girls. He grew up entirely during the last part of struggle for independence, and he was in early thirties when India got independence. He was full of enthusiasm for the new independent India, and he was politically very active. It would have been smart for him to have joined the Congress party, which was very popular because of its identification with the successful independence movement. The Congress party could also boast of many leaders that were very highly respected throughout India. This was the party of Mahatma Gandhi, Father of India; Jawahar Lal Nehru, the first and one of the longest serving prime ministers of India; and a number of stalwarts of the Indian Independence struggle. However, he disagreed with policies and methods of the Congress party, So, he joined a minority party called the Swatantra party. He ran on very little popular and financial support, and he lost. This loss also made him poorer, and his family suffered a lot. But he believed in what he was doing and kept working at politics unsuccessfully for a long time.

GROWING UP IN MUMBAI, INDIA IN 1940S, '50S AND '60S

As part of his political activity, he founded and edited a small newspaper that he printed himself. The mission of this Hindi newspaper was to expose misdeeds of politicians in power. This was not a smart choice. He did not make much money because the circulation of the newspaper was low. Furthermore, he made a lot of powerful enemies. He still needed to make money. With his newly acquired knowledge of the printing process and using printing equipment that he already had acquired, he started a business making school notebooks and printing textbooks. His children joined him as they came of age, and eventually turned this business into a successful business that nicely supported their family.

Surendra was the oldest child and he was my age. We got to know each other very well. Whenever I was in Ratlam, he was always with me. He went with me wherever I went. And I went with him wherever he went.

When Prime Minister Indira Gandhi felt threatened and declared an emergency from 1975 to 1977 due to a growing revolt against her government led by Dr. Ram Manohar Lohia and Jayaprakash Narain, a large number of opposition leaders were arrested under Emergency powers and a new special ordinance called MISA, Maintenance of Internal Security Act, which gave Prime Minister of India Mrs. Indira Gandhi and all law enforcement agencies almost unlimited powers to detain anyone on suspicion alone without formal charges or trial. Opposition leaders were detained and tortured for an indefinite period, without any formal charges. Chhote Mamaji was also arrested. But he was released after he had spent a few days in jail.

Mrs. Gandhi's party lost the next election badly when the Indian people totally rejected her abuse of power. Many of these opposition political leaders came to power and formed a non-congress central government for the first time thirty years after Independence. Chhote Mamaji did not get included even in this wave.

Whenever I was in Ratlam, I always visited Masaji and his son, Bhanwar, who was a few years older than me. Masaji dealt in gold jewelry. It was not common for most poor to middle class persons to have bank accounts. For many, the complicated process of depositing and withdrawing from Indian banks was not appealing. Also, many banks failed, and when they did, people lost money. In addition, preparing tax returns and paying income taxes did not appeal to them. Things are changing, but cash was and still is used a lot. A great part of economy of India is done in what is called black money—money on which no tax is paid on sales and profit generated. It is risky to deposit such money in a bank since it could be scrutinized for tax evasion. For these and many other reasons, profits and savings are invested in gold, jewelry, precious stones, real estate, and other tangible properties. Another common way of keeping black money was to loan it at high interest rates to reliable borrowers- friends, businesses, etc. However, these transactions were done without any enforceable paperwork. My Masaji, in addition to trading in jewelry, was also a lender at a high interest rate, much like a pawn shop, with gold jewelry deposited by a borrower acting as collateral. My friend Bhanwar also worked in the shop. It was interesting to see him handle so much cash. I had never seen so much money. All I had handled was my small pocket money.

Traditionally, Indians tended to have large families at that time. And families were close knit. I had over ten direct uncles and aunts and we were all close to each other. There were many more distant relatives, including uncles and aunts, and many of them were also very close to us. Then there were friends of the family who were so close that we considered them family and called them uncles and aunts and other relations depending on age. All of these uncles, aunts etc., had an average of five to seven children. Every one of these children was considered brother or sister. So, you can see an extended family that was close to each other numbering in several hundreds.

Everyone was invited to weddings, and most of them accepted invitations and attended these weddings and came with their entire family. There was no RSVP restriction on numbers. I enjoyed meeting as many of them as possible. Some I only saw and smiled, while others I had small talk. But there were many that I got to know better and we talked at great length and I tried to learn more details about their life. Once we parted, we generally did not communicate until the next wedding. In those days, a phone was not easily available, and it was expensive to use, so children calling each other was not common at all. Once in a while we wrote letters. Or at the next wedding, you picked up where you left off. Knowing all those cousins was one of the greatest joys of my life.

My Chhoti Mamiji and my Dadiji were very fond of me. We did not talk a lot. Both spoke in Marwari to me. I replied in Hindi. Both were very happy when I was around. And I was very happy to be with them. Both asked me questions about my school, my friends, my hobbies, and anything that they could think of, just so that they could hear me talk. Their faces beamed whenever they saw me. That made me happy, too.

They showed their love by cooking all kinds of food, some that they knew I loved. And some that they knew I had never tried, but that I should try.

They also taught me to eat properly by taking only the food that I was going to finish and no more. Goddess of food, Annapurna, is revered highly. Wasting leftover food is considered an insult to her. If you did that there is a risk that you could yourself be deprived of food by her as punishment. They, along with Baiji, always reminded me that it would be unconscionable to waste food while so many people in India and in the world could not get enough to eat. They taught me to respect food and always finish what I took. They also made sure that the amount of food prepared by them was just the right amount. They did not want to cook too much; if food was left over, they would have to stuff themselves to finish it, which they sometimes ended up doing. By the same token, if someone ate more than expected, family members that followed would sacrifice and eat less. We did not complain. We knew the reason for shortage, and we accepted it. I also observed that many adults, after finishing every bit of food in their thali, would pour water into thali, swirl it, and drink the water at the end of their meal so that not even a speck was wasted. I think it is a great idea, but I could never bring myself to do so.

I noted one thing peculiar about the food of Madhya Pradesh. The single most common ingredient was Besan, gram flour, similar to flour made from chickpeas or garbanzos. It is very beneficial since it is high in carbohydrates, which provides badly-needed calories and protein. These nutritious components are important, particularly for the vegetarian Indian population that traditionally starved. Baiji also made these foods. But she also added Gujarati and South Indian dishes that she learned from her neighbors to her recipe collection.

My favorites among them are listed below. Most of them are served with hot chutney made of coriander and hot green peppers, and sweet chutney made of tamarind or dates.

Bada or Bonda-besan fried balls approximately 2 inches in diameter stuffed with spicy and salty mashed potatoes.

Bhajiye/Pakora-similar to Bada, except smaller and shaped around the ingredients such as onions, hot green peppers, bananas, hot green peppers, sweet peppers, etc.

Sev-small spicy fried noodles or sticks in varying thicknesses made from besan and squeezed through various sieves of different size holes.

Besan-cooked by itself with spices and water and consumed as a substitute for vegetables to eat roti or paratha with.

Boondi Raita-liquified yogurt with boondi – round ¼" diameter fried besan balls.

Kadhi-boiled buttermilk with besan. Fried okra and pakoras are added to make additional variations of kadhi.

Kachori-a ball of flour in various sizes and shapes is filled with ingredient such as besan, moong, urad daal spices, and then fried. Varieties include onion-filled kachori and even mawa-filled kachoris.

Gatte ki Sabzi-made from boiled and hardened besan rolls sliced into smaller pieces and made in dry or soupy versions.

Pitod-besan rolls.

Chidla (Pudle)-thin pancakes made from besan.

Paapad-thin salty, spicy, and crispy, made from a variety of lentils.

Madhya Pradesh Style Vegetarian Thali

All these items were made with a lot of ghee and with lot of salt and red pepper. All kadhi and vegetables made with ghol/rassa had a very inviting floating melted ghee of dark red color of red pepper.

Sweets-while I preferred salty snacks over sweets, there were also many sweet dishes made from besan. My favorite was besan barfi, besan laddu, motichoor laddu, sohan paapdi (pateesa), sweet boondi, and mohan thaal. You had to have a lot of ghee (butter made from cow's milk) and a lot of sugar to really enjoy them.

Please don't get me wrong, there were also many items made without besan. I remember a Rajasthani Thali/plate. There were a dozen items in the thali. In addition to dishes made from besan, some of non-besan dishes were baati churma, lasan ki chutney, mirchi ka kutaa, ker sangri, odave, pachkutaa, raab, lapsi, aate ka halwa, baajre ka sogra, and a variety of papad, chutney, and pickles. In addition, there were also non-besan sweets such as mawa ki kachori, makhanbada, malpua, ghevar, gulab jamun, rasgulla, ras malai. South Indian dishes added by Baiji were idli sambhar and masala dosa. Gujarati dishes added by Baiji were ganthia, khaman and dhokla.

CHAPTER **14**

Romance and Bollywood

WE LIVED IN the chawl, where in our building alone we had an estimated two hundred to three hundred children or young adults going to school or college. So, there was plenty of choices as to who you wanted to associate with and who you chose for friendship.

As we were growing up, we were taught by our parents and all older relatives to view all girls as sisters. In general, we primarily looked at their faces as we interacted with them. It would be embarrassing if you were caught looking below the face. I did observe this rule for girls in our building and in my school, until I got older when I started cheating. Girls who were not from our building, or from my school, or whom we did not know, because of our parents or relatives, could be admired in more complete ways, I assumed. However, it was difficult to get to know them because no one would introduce us or arrange any activity with them.

During primary school, even though it was a co-educational school, I don't remember much in the way of interaction with girls. We viewed them as competition in studies, games, or sports.

When I entered high school at age eleven, and until I graduated from high school at age sixteen, I definitely was interested in girls.

ROMANCE AND BOLLYWOOD

However, it was an all-boys school. So, there were very few opportunities to meet girls.

When I entered St. Xavier's College for two years to study sciences, I did have lot of pretty girls in my class and college. I was able to talk to them briefly, here and there. But it was out of the question to express any feelings of attraction. That did not stop me from singing romantic songs loudly so girls could hear and hopefully figure me out. However, to be on the safe side, I still pretended that I was singing to enjoy the melody, and there was no other purpose. I certainly would not have admitted that it certainly was to get the girls' attention!

Almost every Indian movie, no matter in what language, is a musical movie, with an average of six to eight songs of three to six minutes duration each. Most of the songs are romantic songs, expressing love of a boy for a girl. So, there was plenty of choice of popular songs to choose from to suit a particular mood and girl.

After two years at St. Xavier's College, I joined engineering college at Victoria Jubilee Technical Institute, abbreviated VJTI, to continue my studies. Indians love abbreviations. Sometimes they are even 8 to 10 letters long. There were approximately ten girls during the three years I was in engineering. I was more aggressive than most of my classmates. I approached the girls and talked much more freely with them. But there was no romantic interest. Certainly not from their side.

Bombay, the financial center of India, is a large metropolis. In our building, we had people from various parts of India. Bombay is in the Maharashtra State. Maharashtra State has its own languages, primary one being Marathi. However, most people who migrated to Bombay often retained to a great extent their language and cultural heritage of the original place where their ancestors came from, even after spending a number of years and even generations beyond their original place.

GROWING UP IN MUMBAI, INDIA IN 1940S, '50S AND '60S

Chandu was my neighbor and closest friend. Chandu's father, who was from Uttar Pradesh, spoke with his visiting relatives from Uttar Pradesh in a particular form of Hindi, often referred to as Bhaiya, meaning brother. Most people in Bombay at that time knew this particular variation of Hindi, mainly from contacts with a person selling milk door to door, or with scheduled home delivery of milk, or in established milk shops. Bhaiyas dominated the milk business. The price of milk, a very perishable quantity, fluctuated a lot. Signs similar to prices displayed by gas stations, except written in chalk on a slate board, announced price fluctuations as they happened. In case of shortage, or even otherwise, water was often added to milk to increase profit.

Chandu's family lived in the last apartment on our floor on our side of the stairs. There was only one neighbor between his apartment and ours.

Chandu's mother had grown up in Gujarat. She spoke Gujarati, even though she was also originally from Uttar Pradesh. Because of Chandu's mother, the entire family spoke Gujarati and the kids went to Gujarati Schools.

Chandu's dad had a job as the manager at Bandra Talkies Movie Theater. Movie tickets started at five annas (one anna was 1/16th of a rupee). Our parents did not want to spend that much money. They also were not in favor of corrupting our young minds with movies, which were generally made for mature adults who could spend money. So, we were not allowed to see many movies. Chandu, however, could see as many movies as he wanted at his dad's theater for free. Also, any restrictions imposed by the censor board related to age did not apply to him. He could also get free access to other theaters because of his dad's contacts. He therefore rarely missed any movies, and often saw many movies many times.

A Still from Andaz (1949) Starring from Left Nargis, Raj Kapoor, Dilip Kumar

Because of this, he learned, by heart, many dialogues in the movies. He also saw how much money and fame successful heroes had. He believed that he had nearly as much talent as these heroes had. So Chandu aspired to become a hero in a movie. He was handsome. He was the right age for an acting career. He developed acting talents with his own efforts. He learned to sing. He wrote scripts. He produced and directed small plays with other kids. He performed them free for residents of our building.

In 1958, he saw the movie Madhumati. I was fifteen, he was seventeen. This movie played such an important role in Chandu's life that I am going to tell you more about this movie.

In order to fully enjoy Madhumati, you have to accept the theory of reincarnation, an almost endless cycle of birth and death. You must also believe in ghosts. If you don't believe, you must at least allow reincarnation and ghosts as a possibility, for at least three hours, while you are watching the movie. Fortunately for the makers of Madhumati, 80% of Indians are Hindus. An additional 5% are Buddhists, Jains, Sikhs, or other similar religions, who also believe in reincarnation. The remaining 15% Indians are Jewish, Muslims, Christians, and similar religious faiths who also believe in the afterlife and can certainly accept reincarnation temporarily so they can enjoy the movie. Their family connection with Indians of other faith over generations, and circumstances of being surrounded by a Hindu majority may make it easy for them to accept reincarnation as a possibility.

There is one problem though for the Bollywood movie maker. All religions—Hindus, Buddhists, Jains, and Sikhs, that believe in reincarnation, also generally believe that all living beings have a soul, and that the soul is eternal and never dies. These theories make it very clear that when a human being dies, they are not necessarily coming back as a human being. Their karma and mercy of God will determine what species their soul will occupy in the next life. But Bollywood does not care for this unnecessary detail. In Bollywood, human beings come back as human beings – period.

Hindus also believe that unless death is honored by a proper religious ceremony including rituals requiring feeding priests and guests, the soul does not enter another body and wanders in distress as a ghost. There may be other reasons why a soul may become ghost. Bollywood uses revenge, love, friendship, etc., as reasons why a soul may become ghost and remain so for some time. With this brief summary of some of the beliefs that Bollywood uses to entertain you, let us see if we can follow the story of Madhumati.

The hero is played by Dilip Kumar who has two roles— Devendra in present life, and Anand in previous life. The heroine, Madhumati, is played by Vyjayantimala who is killed and becomes a ghost, also played by Vyjayantimala. Vyjyantimala has two additional roles. She is Madhavi, who is trying to help Anand avenge Madhumati's murder. She is also Radha, who only shows up for about a minute at the end of the movie. The third most important character is the villain Zamindar Ugranarayan, played by Pran, who does not age in the movie at all while others die and are reborn. Now the story.

Devendra (Dilip Kumar) is being driven in car on a rainy dark night on a narrow curvy road without shoulders in a hilly terrain. Devendra is rushing anxiously to pick up his wife, Radha, and their newborn daughter. Radha is returning from her parent's home where she had gone for delivery. The train station where she will arrive is about 20 miles away. Because of heavy rain and strong winds, a big tree had fallen onto the road and blocked Devendra from continuing on his trip to the train station. Devendra is forced to take refuge in a big, abandoned mansion nearby.

Devendra feels he is familiar with this mansion, even though he had never been there. He seems to remember something that happened in the mansion. He questions why a picture, which he feels like used to hang on the wall, is missing. He asks if there is a lake nearby. He seems to remember the lake, even though he had never been in the area before. With all these vague memories rushing into his mind, Devendra feels that he has some connection with this house.

He hears the screaming of a woman. He hears the sound of ghungru, anklets with bells. He hears footsteps. He decides to investigate. He hears screaming again. He hears a woman crying. A picture falls off the wall. He remembers painting that picture of a rich zamindar. He hears someone calling out "Babuji". All this helps to bring his faded memory back very clearly.

Now he knows that he was in this place in his previous life. His name then was Anand. He painted this picture when he worked as a manager for Ugranarayan's timber estate called Sundar Estate.

The movie flashes back to tell us what happened in chronological order from the beginning.

Anand sings a song "Suhana Safar" (beautiful trip) as he is on his trip to start his new job as manager of Timber Estate. Even though the movie is black and white, the scenery is fantastic, with mountains, rivers, and a thick forest. Anand is all alone, not a person in camera's site. He is looking forward to all that his future might hold for him.

Just as the song ends, a village woman spots him getting close to the edge of a cliff, carelessly and absentmindedly, while he was looking upward at sky and singing and dancing. She saves his life by keeping him from falling off. He thanks her. The hero has been introduced to the heroine.

Anand arrives and checks into his new residence. There, he keeps hearing "Aaja Re Pardesi" (welcome foreigner). He goes out to investigate who is singing and enters an area marked by a sign that reads "Forbidden Territory". There is enmity between people inside "Forbidden Territory" and those outside it. But Anand is not afraid. He disregards the sign and enters the forbidden territory. He hears the song "aaja Re" again and follows the sound and locates a woman (Vyjayantimala) singing the song at the bottom of a fall. In the song, she says she is missing his Pardesi with whom she has had a love relationship over many lives. Anand chases her. He discovers that it is the woman who saved him and gets a brief introduction and finds out that her name is Madhumati, or Madhu for short.

Anand wants to meet Madhu again. He decides to go to a fair/mela to see if he can meet her again. He finds her dancing and singing another love song "Julmi sang AankhLadi" (my eyes met the torturer). Translating Indian songs is impossible.

Anand saves the child from getting trampled under a horse being ridden by Ugranarayan. Madhu is impressed by this brave act of a caring man. She is no longer afraid of him. Actually, she likes him. She approaches him to ask him to come to her home so she can provide medical care for his hand that he injured while helping the child. Now they are in love. After this visit, they meet regularly in the evening at a set time of 8 p.m., and at a set place, which is marked by a specific rock and a specific tree.

Anand, whom Madhu calls Babuji, also happens to be a good painter. He makes a painting of Madhu. Ugranarayan sees the painting and inquires as to whose picture it is. Anand suspects that Ugranarayan has bad intentions, so he lies and says the painting is just from his imagination. But Ugranarayan does not buy the lie and finds out that the beautiful lady in painting was none other than Madhumati.

Ujranarayan sends Anand away for an assignment to get him out of the way. Anand does not inform Madhu, and Madhu, unaware of his trip, goes to their meeting place at the usual time. Anand does not show up. While she waits, she dances and sings "ghadi mera dil dhadake" (my heart is beating every second again and again). But instead of Anand, Ugranarayan shows up and attempts to molest her. She is able to get away, even though she is running barefoot, and he is chasing her on a horse. She knows the short cuts and he does not.

Anand and Madhu continue to meet. Anand sings "dil tadap ke keh rahahai" (my aching heart is saying…). Ugranarayan spots them and tries to catch up, but Anand and Madhu get away from him and end up in a village mela. She sings "chadh gayo paapi bicchua" (a cruel spider is going up my body).

GROWING UP IN MUMBAI, INDIA IN 1940S, '50S AND '60S

One day Madhu's father catches her trying to leave at night to meet Anand. He is against her plan to go out, particularly because of long term enmity of his people within the forbidden territory with Ugranayrayan's Estate. He is against Madhu's involvement with Anand who is a foreigner and who works for Ugranarayan. He orders her to stay home and never meet Anand again.

When Madhu does not show up for their evening time together, under heavy rain and storm, Anand decides to go to Madhu's home to investigate why she did not come. Madhu apologizes and promises that she will never leave him again, even after death. In the future, she promises she was going to ignore her father's orders. Her father wakes up. Anand tries to run away to avoid a scene. Her father catches up with Anand. Her father realizes that Madhu is in love with Anand, so he asks if Anand is also in love with Madhu and if he will marry her. Anand says yes. Her father now happily accepts their decision to get married. Her father goes to the bazaar to shop for their planned wedding.

In the meantime, Ugranarayan wants Madhu to satisfy his lust. Ugranarayan asks Anand to go on a business trip again to get him out of way. Anand goes to Madhu to inform her of his trip. Madhu is afraid of Ugranarayan and is concerned about her safety. She asks Anand to go to temple with her to receive blessings to keep both of them safe. They go the temple. At the temple, Devendra goes beyond obtaining a blessing, and applies Sindur to Madhu's maang/hair part line, in a symbolic marriage, with God as their witness.

Anand leaves for his trip. Ugranarayan sends a message to Madhu that Anand had been hurt in an accident, and tricks Madhu into coming to his mansion. Madhu arrives and he attempts to rape her. She runs away but falls off the protective guard wall and dies.

Madhu's father returns from his shopping trip, and Anand returns from his business trip. Both discover that Madhu is missing. Both start looking for her. Madhu's father spots Madhu's dupatta and concludes that she is dead. Anand refuses to accept that conclusion. He runs around looking for her repeatedly shouting "Madhu". He had done so many times before and always found her. But this time, he does not find her.

Anand eventually realizes that Ugranarayan had tricked him into leaving. Anand decides to see Ugranarayan to find out what is going on. Ugranarayan's men overpower him and badly injure him. He recovers physically but is unwilling to accept that Madhu is dead.

He does not have Madhu, but he does have the painting of Madhu that he had made earlier. The painting is all he has of her, and he constantly keeps it with him, and stares at the painting all the time.

Madhavi, a look-alike of Madhumati, had come to a nearby town to perform in a charity show. Madhu, her police inspector brother, and his friends were hunting for fun. Anand, who happens to be passing, sees Madhavi, confuses her for Madhu, and tries to talk to her. The police inspector and his friends are not too happy about that. They ask Anand to leave Madhavi alone. He apologizes and leaves. But he forgets to take his painting. Anand is heartbroken that Madhavi was not Madhu and sings "tute huye khwabon ne humko ye sikhaya hai" (my broken dreams have taught me this...).

Madhavi spots the painting that Anand forgot. She realizes that Anand had made an honest mistake in confusing Madhavi for the lady in the painting. She concludes Anand was not a bad person. Madhavi needs to apologize and return the painting to Anand. She locates Anand and goes to his home. While she is waiting for Anand to return, Anand's servant tells her about Anand and Madhu's love.

The clock strikes 8 p.m. and Anand again hears Madhu's voice and part of her song "aajare" and follows Madhu's voice. Anand finds Madhu's ghost who tells Anand that Ugranarayan had attempted to rape her and she died by falling off the edge of the terrace roof as she was running away to avoid rape.

Anand meets Madhavi's brother, the police inspector, who might be able to help. He complains and asks why Ugranarayan had not been charged and tried already. The police inspector explains that even though Ugranarayan had a bad reputation, they had no prosecutable case against him without proof.

The police inspector invites Anand to the show, in which Madhavi was to perform. She sings and dances "Bichhua" (Spider). Anand realizes that Madhav not only looks like Madhu, but she also had singing and acting talents, and comes up with a plan to obtain Ugranarayan's confession. If Madhavi dresses up as Madhu's ghost and terrifies Ugranarayan, he will confess to his attempted rape. Anand presents his plan to Madhavi and the police inspector. They decide to go along with his idea, and they make a detailed plan.

Anand goes to Ugranarayan and tells him that he is unhappy, unemployed, and broke, and would like to earn money to pay for his trip back to his home. He requests if he can paint Ugranarayan's picture to earn money for his trip. Ugranarayan agrees to a 7:30 p.m. appointment the next day.

Anand arrives at the appointment at Ugranarayan's mansion and begins painting. He talks and tries to scare Ugranarayan in preparation for Madhavi's arrival as Madhu's ghost.

At 8 p.m. sharp, a strong wind starts blowing through the open windows, swinging a big chandelier; the lights go out and Madhu's ghost appears. Madhu's ghost scares Ugranarayan and asks him to confess, and he confesses to attempted rape which led to Madhu's death. The police in the meantime had been hiding close by and hear his confession and arrest Ugranarayan to start legal proceedings.

Anand thanks Madhavi for playing Madhu's ghost and helping him. But he suddenly realizes that Madhavi in her performance mentioned events that Madhavi could not have known. He is confused.

Just then Madhavi 's car pulls in. She runs in apologizing for being late. For a moment, Madhavi and Madhu's ghost both are in the room at the same time. Anand knew that it wasn't Madhavi, but Madhu's ghost, who had helped convince Ugranarayan to confess to his crime.

Madhu's ghost disappears, leaving Madhavi alone in the room. Anand turns to find Madhu's ghost. He hears Madhu's voice calling him out. He hears Madhu singing "Aajare pardesi". Anand sees Madhu's ghost in open space beyond the edge of the mansion roof. Anand follows the voice and he falls off the edge of the terrace roof to his death and joins Madhu's ghost.

Devendra just finished the story of Anand and Madhumati. He knew that Anand and Madhumati had to meet again in a new life because of their true love and their unfinished love story. Devendra suddenly realizes that he is Anand and that is why he knew the full story of Anand.

The driver comes in and informs Devendra that the storm has passed, that the tree had been removed, and that they can go to the station. However, there was bad news—the train that Radha was going to arrive in had been in an accident. There were a lot of injuries and some deaths. Devendra rushes to the station and is happy to find Radha safe. He immediately knows that Radha is none other than the reborn Madhumati. He tells Radha that he did not want to lose her again. Radha looks at him and wonders why he said "again". Devendra explains to the audience that he got Madhumati back as Radha and that they will be together life after life. A baby is heard crying. Devendra spots his newborn daughter. He smiles.

The movie ends.

Chandu either thought that his life paralleled Anand's life, or he made himself and others believe that it was so to further his acting career aspiration. Chandu also fell in love with a girl, Kumari, who lived in an apartment near the stairs. If you shouted from the stair area, she could hear you.

Girls in India were forbidden from making friends with boys. The only relationship acceptable to society before marriage between a boy and girl was a brother and sister relationship. Even friendship was not acceptable. The slightest hint that a girl may have had romantic interest in any boy, or a boy ever had any romantic interest in her, could ruin her chance of finding a good marriage partner in an arranged marriage. Her parents had the sole responsibility to arrange the marriage. When parents are trying to arrange a marriage, it is so easy to pass up a person even for a minor flaw. There is no attachment at that point with the person—only a calculated evaluation.

Kumari probably was not interested in Chandu. Even if she was, she could not and would not have shown it. And she did not. But that did not stop Chandu. Chandu wrote poetry and delivered it to her through messengers. Chandu would sing songs from Madhumati loudly near her apartment. Chandu would scream "Madhu! Madhu!!" in a good echo-filled imitation of Dilip Kumar in the Madhumati Movie. But unlike in the movie, there was no response from Kumari. He tried again and again and kept failing to get her love, her friendship, or even her attention.

But Chandu was just not going to give up. Chandu continued singing songs near Kumari's apartment at all hours of day. One song from

Madhumati that he sang the most was "Suhana Safar aur ye mausam-hasin..." (pleasant trip, and beautiful day... may be time to get lost in love...), if he was in good mood. If he was in sad mood, he would sing, "Tute hue Khwabaun ne humko ye sikhaya hai" (broken dreams have taught me ...). He would scream romantic dialogues from the movie, hoping that she would hear them. I am sure Kumari heard the songs and dialogues sometime, but she did not, or maybe could not, show any interest in him. Heartbroken, after about a year, he gave up. He recovered from this episode; I think.

CHAPTER 15

Dream of Becoming a Bollywood Star

CHANDU WAS TOTALLY into Hindi Movies. He enjoyed them, and he really wanted to somehow succeed in the movie industry. He and I were closest of friends, and we spent a lot of time together. He wanted me to do everything that he did. He wanted me to try to succeed in the movie business, too. He talked to me a lot about it, hoping to ignite a passion in me for movie industry. He got professional pictures taken of him. He saw a lot of movies, many of them over and over again. He memorized the dialogues and delivered them almost perfectly. He acted out scenes, pretending as if he was performing before a movie camera. He looked for every opportunity to improve his skill, and thereby improve his chances, to become a successful hero. If he succeeded, he would be rich, famous, and highly-admired. He wanted the same for me.

Chandu was able to arrange visits to film studios to watch filmmaking. He took me sometimes. One time, I saw Meena Kumari as she got out of her car to go to the studio. The experience was unnerving. She was really short, even though she looked normal height on screen. She had rough skin, and she did not look as pretty as she appeared on screen. I found out how crude and fake looking the sets

DREAM OF BECOMING A BOLLYWOOD STAR

were. Her make-up also was too dark. I found out how a typical take required an actor to speak just a few sentences at a time. They got so many chances and they got it right only after so many re-takes. I was disappointed with this make-believe world and it diminished my interest.

Chandu, on the contrary, was inspired by the actors. He wanted to become one also. He wrote plays. He played the main part on many plays. I also acted in his dramas in smaller roles. We made hand-drawn posters, and pasted them on walls, on all the floors, near stair area announcing our upcoming drama. The admission was free; the program was in evening. We had a huge audience; several hundred sometimes. Ropes were stretched, and white bed sheets were hung on them with wooden clips to form the curtains. Dramas were about hour and half long, with several breaks for change of costume. I was just having fun, but for Chandu, this was serious business.

Chandu entered a contest conducted by a program on Radio Ceylon. The chocolate company Cadbury sponsored the program. It was called Cadbury Phulwari, in which talented children entered a singing competition, similar to American Idol, except this was on radio. He asked me to join him and try out for the contest also. So, I did. We were called together for the preliminary selection audition. It was really hard to learn songs in those days. We did not have access to printed lyrics and heard songs only when they were played on radio, and if we happened to be listening. There were just a few programs that I was able to hear that played songs. Half an hour or so on Radio Ceylon, and half an hour or so on Vividh Bharati. You were lucky if the song was repeated within a few days.

My friend Chandu was luckier. He owned a vinyl record player, made by RCA, under a trade name His Master's Voice, with a picture of a cute pet dog as their trademark. This record player was mechanically driven. You turned a handle for a minute or so to store energy in a spring. When you turned the record player on, the spring released enough energy to turn the table to complete a 3-minute song,

Songs were recorded on a vinyl disk, with one song on each face, or one of two parts of a longer song on each side. You had to wait until the disc started spinning, and then quickly set the pin down in the first line of grooves carefully or you would scratch the record. If you wanted to replay even a part of it, it was best to start from beginning and wait for your portion of interest. There was no fast-forward, rewind, or pause. If a record was scratched, it would get stuck in the same groove and keep playing the same three words over and over again.

Regardless, we practiced as best as we could. We went to Radio Ceylon recording studios located in downtown Bombay. While we were waiting, we saw Amin Sayani, one of the most popular radio announcers of that time, go by. We auditioned. Unfortunately, both of us failed in the very first audition. This was a small setback, but this was not going to keep Chandu from trying more. For me it was fun to go and try. But I gave up.

Navratri is a festival that is celebrated with great zeal and enthusiasm. Gujaratis celebrate it most enthusiastically. It is celebrated not only all over India, but almost everywhere in the world where Indians live. During these nine nights, nine forms of Durga/Shakti/Devi are worshiped.

A huge public celebration in Shivaji Park, close to our home, was arranged by the Gujarati community. It was attended by several thousand people on each of the nine days of Navaratri celebration. Program started with Garba. Participants were women wearing traditional Gujarati chania choli or Saris. This was followed by Raas, in which both men and women participated. They used raas dandia/sticks. The program ended with Durga puja and Prasad. On the final day, the garba and raas were shortened. A fancy-dress competition was held. The competition was broken into different age groups. Chandu entered the contest. I also entered the contest with Chandu's encouragement and help.

I was in eleventh grade, my last year in high school. Chandu came up with my character for the contest. I was going to be Habshi, an African tribesman. I was painted black on all my exposed skin with black shoe polish. I wore a skirt made out of coconut tree leaves. I wore a necklace made out milk bottle aluminum foil caps, which were shiny silver on the inside, and blue with the Aarey Milk Colony whole milk logo on the outside. Caps were turned so the audience saw silver circles. A movie song where there was a chorus that went like this "Haiya ho Haiya Haiyaho", imitating an African musical rhythm, was played. I entered the stage jumping around to the beat of the song. I had fun, and surprisingly I won second prize—that too in my first attempt!

Looking back, I am ashamed that I treated another racial group with contempt without any justification. When you see others do something, you assume that it is normal, and you imitate without much thought. It was wrong of me to make fun of an entire race without making any effort to know them better. I am ashamed and I hope that I never do it again.

The following year, no one had to convince me. I entered again. Now I was in my first year of science college at St. Xavier's College. Again, with Chandu's suggestion, I chose to become a statue of Sai Baba. Sai Baba was born Muslim and remained Muslim throughout his life. But he admired and had great faith in Hindu gods also. He is worshipped by Indians of many faiths including Hindu and Muslim. Sai Baba taught a moral code of love, forgiveness, charity, contentment, inner peace, and devotion to God and a guru/teacher. He spoke out against discrimination based on religion, sex, or caste. His teachings combined elements of Hinduism and Islam.

GROWING UP IN MUMBAI, INDIA IN 1940S, '50S AND '60S

The most common murty/statue of Sai Baba in his temple is made of white marble showing him in a sitting position on a platform with one leg crossed over the other knee. I was going to be Saibaba as shown in these Murtis. I was painted white with tennis shoe white chalk polish to look like white marble. I put on a beard that resembled Sai Baba's beard. I wrapped a white dhoti all around me in his manner. I sat on a stool, with one leg crossed over my other knee, just like his statue. I put on a look with a slight smile of contentment. I thought I would easily win, but I was disappointed that I did not win any prize this time. The competition was perhaps too good. And my beginner's luck had run out.

There was no government-funded or privately-funded safety net in old age for most people. Provident funds, though significant in current value of rupees, would amount to very little after the value was reduced due to inflation to take care of you in your old age. Daughters would move into their husband's home at marriage. Their fiscal responsibility for the house would generally end with marriage. Sons were the only ones who would be around, and they were the ones who would take care of the parents in old age. The story of Shravan Kumar is told repeatedly by every Indian parent to teach their sons of their obligation to take care of their parents in their old age.

Shravan Kumar's blind and old parents expressed a desire to go to holiest river Ganga for a pilgrimage. Shravan, who was very obedient and would do anything for his parents, immediately proceeded to fulfill their desire. He hung two baskets from a stout wooden pole and carried the pole on his shoulder with one parent sitting in each basket. On their way, they stopped at a pond to take a break. Shravan set his parents down under a tree. His parents were thirsty, so Shravan went to fetch water for them. He dipped his brass jar/lota into water to fill it.

King Dashrath, father of the Lord Rama was hunting nearby. Dashrath, confused the sounds made by Shravan as he was filling his lota with water, thought the sounds were made by deer that were drinking water from the pond. He aimed his arrow toward the source of sound, impulsively without visually checking. When Dashrath went to claim his hunt, to his horror, he discovered that he had hit Shravan.

Shravan was bleeding and dying. Even as he was dying, he was more concerned about his parent's thirst, and he requested Dashrath to deliver water to his parents first and not worry about him. Dashrath proceeded to do so. Dasharath took care of his parents' thirst and returned as quickly as he could to help Shravan, but Shravan had already bled to death. Dashrath went back to Shravan's parents to give them the sad news of Shravan's death.

Shravan's parents were distraught and angry. Dashrath apologized profusely and explained that he made an innocent mistake. But Shravan's parents were not going to forgive him. They gave him a shrap/curse that he would also suffer a similar grief from a similar loss.

This is the reason Dashrath had to suffer the separation from his oldest son, Lord Rama, whom he had chosen to succeed him. Dasharath was forced by one of his wives, Kaikeyi, to ask Rama to leave his kingdom and not return for fourteen years. Rama spent fourteen years in the forest accompanied by his wife, Sita, and his younger brother, Lakshman. Dashrath died before Lord Rama returned due to grief from this separation.

Every Hindu parent tells their children Shravan's story to teach them that they should grow up to be like Shravan and take care of them in their old age. The story is effective. Until recently, very few children thought of nursing homes or retirement homes as the proper place for their parents, even if it takes great sacrifice on the children's part. Shravan sacrificed his life so his parents could get a drink of water.

So any sacrifice that the children may have to do would be nothing compared to what Shravan did for his parents.

Like every Indian parent, Chandu's parents also tried to teach him to think of his future, and also to prepare himself, so that he can take care of his parents in their old age. But Chandu had a dream, and he wanted to pursue it at all costs. I was in college studying to be an engineer. I was good in studies. I felt confident that I would succeed in my life. I lost my interest in movies, drama, and acting. I knew that I did not have the needed talent to make a career of it. I did keep my interest in singing, but only as a hobby. I did take some lessons later in harmonium, but I didn't have the patience to continue for too long.

Chandu, on the other hand, tried and kept trying. He did not succeed. He had neglected his studies, and he had not acquired any marketable skills. So, he struggled financially in his life for a long time, and he was not able to fulfill his duty to take care of his family very well.

Chandu's family went through a series of crises. Chandu's mother, they called her Ba, had been diagnosed with cancer. Cancer care was very expensive, and the recovery prognosis was not good. Chandu's Ba knew of their financial situation, so she ruled out expensive medical care. Instead, she tried more affordable faith healing and Ayurvedic medicine. These treatments did not work, and her condition deteriorated. She died at just over forty years of age. Chandu's oldest sister, Bindu, who was an excellent singer, tried to fulfill the role of her mother at young age. But the stress was just too much. She suffered a severe mental disorder. One of the side effects was that she could not get up by herself. She started seeing faith healers. In her case, a miracle did take place. She regained her ability to get up suddenly one day. However, the depression of her situation overwhelmed her. She refused to get married, so that she could take care of her younger sisters and brother. She was very frustrated and depressed. A few years later, she committed suicide by jumping from their fourth-floor apartment to the concrete sidewalk on main road below. Chandu's dad also died at an early age from heart attack.

Chandu was able to establish a small silk-screening business with decent income. Even though he went thru some tough time, I am glad he tried to follow his heart. The new business did satisfy his desire to be an artist while providing income for his family.

CHAPTER 16

Neighbors

THERE WERE NINETY apartments in our two building chawl apartment where five hundred tenants live, each with a story worth telling. People from the majority of states in India lived here. They spoke different languages at home. They communicated with each other in their Hindi and/or English as best they could. I was lucky to grow up in such an environment. I am going to describe a few neighbors who were closest to me.

After Chandu, Jayesh, who was probably a year older than me, was my best buddy. Chandu was often busy with his efforts toward a film career. But Jayesh was always available. So, we whiled away our time playing marbles. We also played other games with soda caps and empty cigarette packets as substitute currency. Jayesh was good at these games. I was not. So, I would usually lose. I would incur a large amount of debt. When I could no longer afford to play, he would forgive some of my debt, and we would resume playing. I played cricket and many games with him. Jayesh taught me how to ride a bike. He would rent a bike at my expense for his height. He would teach me and would enjoy riding using my rental time. I did fall off the bike and hurt myself a few times, partly because I was riding a bike too big for me. But I did not mind.

NEIGHBORS

Jayesh and his older brother, Surendra, taught me how to swim. Surendra was very enterprising. He figured out how we could get into one of possibly only two swimming pools in Bombay at that time. A small period was set aside for card-carrying members of the textile labor union. We pretended to be textile union workers and sneaked in during this limited period a number of times while the checkers were distracted. That is where I learned to dive from a springboard and to float on my back and swim using a backstroke. I also learned the butterfly stroke. I never became a good swimmer, but I learned enough that if I fell from a boat, I could survive if help arrived quickly. I was never afraid of water.

Surendra was closer to my older brother Babu's age, and he was mainly Babu's friend. But I respected him, and I talked to him a lot. He would advise me and treat me as his younger brother. He was studying to become an architect. He was very smart; sometimes too smart for his own good. Instead of concentrating fully on his studies and graduating as quickly as possible and pursuing a professional career as an architect, he was busy making money to pay for his education and helping his family. At an early age, he was buying options on land allotted by the government for mining. This was a very risky business, particularly for a young college student like him. Most kids his age would have no clue what the mining options were. And here he was, investing his own hard-earned money into such a risky venture. And he did a lot more: he did consulting in architecture, land survey work, he would buy things at a bargain price in auctions and sell them for profit. Despite all this, he never became rich. But he never stopped trying all his life.

Lallu was the oldest of the Jayesh's four brothers. He did not have much education, but he was very enterprising, though not very successful. Whenever there was a public program that he wanted to attend without buying a ticket, he found some way to get in. One time, he managed to borrow a humongous professional movie camera and carry it on his shoulders. He pretended to be a press photographer and walked in, no questions asked.

India has 30 states and 7 union territories. The current 30 states are Andhra Pradesh, Arunachal Pradesh, Assam, Bihar, Chhattisgarh, Goa, Gujarat, Haryana, Himachal Pradesh, Jammu and Kashmir, Jharkhand, Karnataka, Kerala, Ladakh, Madhya Pradesh, Maharashtra, Manipur, Meghalaya, Mizoram, Nagaland, Odisha, Punjab, Rajasthan, Sikkim, Tamil Nadu, Telangana, Tripura, Uttar Pradesh, Uttar Khand, and West Bengal. The union territories include Chandigarh, Delhi and five island regions near the western and eastern coasts in the south part of India.

Initially, states were drawn up on the basis of language. There were many demands for creating new states for other political reasons. The central government resisted these splits, fearing this could lead to the breakup of the country. However, under immense political pressure, new statehoods were granted based on political and economic aspirations of people and politicians. Five states—Andhra Pradesh, Karnataka, Kerala, Telangana, and Tamil Nadu are often lumped together in the minds of Bombay residents, and many North Indians as one common group referred to as "South Indian". This is offensive to "South Indians". However, the habit has not disappeared. The reason for this lumping together is that North Indians cannot distinguish between these languages.

We had a neighbor who was from "South India". Our neighbor consisted of two brothers and their family. As a kid, I noticed that there was a fixed schedule that determined which brother would sleep outside of their apartment. I wondered why?

Another neighbor was a Konkani family. Alcohol was prohibited in Bombay, but that did not stop the addicted father in this family from drinking. He used to spend most of what little money he made on drinks. In the evening, he would come home drunk. His wife was very unhappy. They would quarrel, and he would beat up his wife. There were perhaps only one or two days a week when this did *not* happen. The oldest daughter was trying to help by earning money. At that time, I did not know how she earned the money. All I noticed was that she would dress up with a lot of makeup and expensive jewelry when she left in the evening. Then one day she disappeared for a long time. Neighbors speculated that she had gone away because either she was going to have a baby, or she was going to have an abortion. She did not come back with a baby.

Another neighbor was a Gujarati family. The oldest son, my friend, was same age as my older brother. His father was very strict and very tight with money. His father would leave very early in the morning and come home very late. He worked really hard to make money. And he was successful. Even so, when my friend needed money, he could not ask his dad. He was too scared of him. His mother also did not have the courage to discuss this matter with his dad. So, she just gave my friend some money that she had hid from his dad. This was same as stealing in the eyes of Jayanti's dad. Whenever he found out he would yell and scream, and sometimes beat my friend and his mother.

My friend and neighbor, Nisha, and her family were relatively prosperous. Contrary to most apartments, which were filled with many children, her family had only two children—she and her brother. Nisha's dad was a successful attorney. He sent Nisha to a private English medium girl's school. She aimed high. She wanted to become a doctor, so she could serve society, and she did. She always guided me and encouraged me to rise above my surroundings and to do something meaningful with my life. She was only two years older than me, but she was so much wiser. She was like a big sister to me. She was an inspiration.

Another neighbor was Kavi. He obtained an admission to study in Germany. I was amazed how he could choose to study in Germany when he did not know German. But he did. He learned the German language, finished his education, and came back to India.

CHAPTER 17

Our Family Grows Rapidly

I WAS THE last child to be born in Madhya Pradesh in my family. All my younger brothers and sisters were born in Bombay, and all the while we lived in Kavarana Mahal.

Babuji was born in or around 1914. Baiji was born in or around 1918. My oldest brother, who died at birth, was born in 1935. My oldest brother Nirmal, Babu, was born in 1937. My oldest sister, Nirmala/Baby, was born in 1939. A sister, born in 1940, who had not been named yet, died at about six months of age.

I, Santosh, was born in 1942. After me, Virendra, my brother, was born in 1944. Sushil, my brother, was born in 1946. Dilip, my brother, was born in 1948. Prakash, my brother, was born in 1950. Urmila/Chhoti, my sister, was born in 1952. Sheela/Doctor, my sister, was born in 1954. Anil/Baba/Gippy, my brother and youngest sibling, was born in 1956. In a span of twenty-one years, twelve babies were born in our family. Seven brothers, three sisters, and two parents—a total of 12 surviving members of our family—lived in our two-room apartment at Kavarana Mahal. Even so, it did not feel crowded. It felt normal.

Celebrating birthdays by throwing parties was not a tradition in our home. Baiji, however, did think about our birthday and would make

GROWING UP IN MUMBAI, INDIA IN 1940S, '50S AND '60S

a special sweet dish like Halwa to mark the day: No balloons, no decorations, no guests, no birthday cards, no cakes. Birthdays were just another normal day for us.

Very few people owned cameras in those days. I certainly did not have one. My dad would occasionally take us to a portrait studio to get our pictures taken.

Author's Family in 1953. Age, at picture, in Parenthesis.
On floor from left: Prakash(3), Dilip(5), Urmila(2), Sushil(7).
On chair from left: Virendra(9), Nirmala(14), Babuji(39), Nirmal(16), Author(11).

One day when I was coming home, I saw a big line running for several blocks going into a shop. I found out that the line was for a professional photographer providing passport size pictures for 50 paisa. I got excited like everyone else waiting in line so I joined them. After about two hours wait in line, I got my flash assisted picture taken. I was told to pick up my picture in a couple of days. When I went

OUR FAMILY GROWS RAPIDLY

back, no new pictures were being taken. There was a line, but much shorter of the people who had come to pick up their pictures. When I got my turn, the photographer tried but failed to find my picture. He promised that he will find the picture later and keep it aside for me to pick up next day. Same thing happened next day and for number of days after that. One day that shop closed. I never got my picture. That was first time someone ripped me off. I was about twelve.

My friend Chandu's family was also growing at a similar rate. When I was seven or eight, I noticed a pattern, and I expressed it loudly to Baiji in the presence of her friends. I said that I noticed that every time Chandu had an addition in his family, very soon, our family also grew. And when we had an addition then Chandu's family followed right along. I wondered out loud, *how come?* Baiji and her friends laughed. I thought it was a brilliant observation, and there was nothing funny about it.

Baiji would often send me to buy and bring mitti or "edible" clay cubes every so often. I did not know then, but pregnant women in India have cravings for this clay and they snack on this clay during pregnancy. I have read that in India and in many African countries, particularly in villages, pregnant women continue to consume it to overcome calcium deficiency. Clay cubes must also taste good to them.

CHAPTER 18

Water Wars

OUR APARTMENT IN Kavarana Mahal had two rooms—a living room and a kitchen and getting ready in morning started with a visit to a 3' x 3' mori area with a high curb to wash your face to wake up more. This area was used as a sink, as a shower, and also as a laundry and utensils/dishes washing area. All these activities needed to be done in this area. These activities were carried out by various members of a family with twelve permanent members and additional temporary guests.

The members spanned a wide range of ages: from toddlers, to college aged children, to women doing housework, to Babuji getting ready to go to work. Everyone's schedule varied and most were on tight schedules. They were students, or needed to go to work, or go to shop for daily supplies. It was necessary, therefore, to come up with a flexible system. It was also necessary to use space outside the apartment for some of these activities.

Brushing and shaving was done outside. We brushed in the exterior corridor. A small brass mug/lota was used to carry sufficient water for the purpose. All the spitting while brushing was done into the street or into the rear interior courtyard. It is possible that if you were walking on the street, that someone could spit on your head. And if they did, you would look up, yell at any person you think is responsible, accept this as part of life, and go on.

A common bathroom activity area consisting of two water taps, a curbed area around the water taps, three toilets, and a connecting walk between the two buildings of our complex, was located on each floor on each end of the buildings. This common bathroom activity area served ten apartments housing approximately eighty persons. Women generally did not hold jobs. They got up early or waited until late morning to visit these common areas. They tried to stay out of the way of men. Anyone who was rushing to go to work or school, or was in severe stomach pain, was given priority as long as they politely requested. But sometimes people could not be polite. Then there was trouble.

Males took showers by pouring water over themselves with a lota/jug using water from tap or from a bucket. Bombay is warm most of the year, so you used cold tap water generally. If it was really cold, then you might add warm water, which was heated in a bamba/heating tank. A bamba is a large brass cylindrical pot heated over coal fired on a rack below. You bathed with pants or shorts on. Occasionally, women also had to bathe in public. If they did, they bathed almost fully dressed. Men changed in public while wrapped in a towel. Women went inside their apartment and changed.

Bombay gets water from several lakes within fifty miles carried by steel pipes up to ten feet in diameter. Two taps on the first floor were directly connected to the municipal water. Taps on all other floors got water from concrete water storage tanks located on two flat terraces on top of the buildings. When the municipal water was on, a pump was started to pump water to the storage tanks. From there it flowed under gravity to all the taps on each floor until it ran dry.

Drinking water, collected from first floor taps, while municipal water was on, was the best water available. It was not as contaminated as water stored in water tanks. It was carried up to four flights sometimes by part time servants, but most often by male or female members of the family.

There was no fixed schedule when municipal water would be turned on. So, people kept an eye on it and informed others by word of mouth. It was collected in brass or steel matkas or buckets. The matka has a small neck to minimize spilling. The shape of the matka and the small neck allowed women to rest it on their waist and hold it with one arm wrapped around the neck. Men did not carry water that way. They carried matkas on their shoulders or on their heads. A piece of sheet was rolled to serve as a cushion for the pot. Most men preferred buckets with round handles. Drinking water was stored in these matkas or buckets, but also in larger brass cylindrical cans with covers.

Water for uses other than drinking was stored in large steel or aluminum drums outside the apartment. You dipped a large steel mug to get water out of them. Women washed clothes and utensils at common taps as much as possible. If tank water ran out, stored water was used.

As soon as Baiji was informed that the municipal water was on, she would send a member of the family immediately to get water from the first-floor tap. She herself also would go many times. Everyone in the two buildings did the same. So, if you did not act fast enough, there would be a lot of people ahead of you. If that happened, you waited a longtime to get your turn. And after all that long wait, it is entirely possible that before you got your turn, water could run out.

Because of this anxiety and all the other tensions of low income, crowded conditions, and hectic Bombay life, people had short tempers and got angry very quickly. They got upset because someone was slow. They got upset because someone was filling too many pots. They got upset because the pots were too big. They got upset when someone cut into the line. Then, you had multiple sides shouting and screaming at each other. Sometimes there were fist fights. Some used this opportunity to settle other scores. Water wars were common.

CHAPTER 19

New Apartment in Andheri

BABUJI HAD BEEN promoted further, and he was ready to get out of the chawls and move into a bigger, nicer place. He selected a place in Andheri West. He rented two apartments side by side with private balconies and did some alteration to make it into one larger and nicer apartment. We moved into this new apartment in 1956.

From Andheri Station, you walked about fifteen minutes on Ghodbander Road. You made a right turn onto a dirt road and walked for about five minutes. On a rainy day, this road would turn into mud and slush. There was no development at that time around our building. The area was totally wooded. It was hard to imagine a place like this in Bombay. We were on the second floor, which was much more convenient for Baiji and Babuji than fourth floor that we had to climb up to in Matunga.

During the day, I saw lot more of nature in Andheri. First, we were in the middle of a forest, which is so rare in Bombay. I saw all kinds of birds and squirrels. Fortunately, there were no animals that would attack. I saw fish, frogs, and turtles swimming in a septic tank under construction, and in big puddles that were everywhere due to the construction in progress. I also saw earthworms on the gravel road.

NEW APARTMENT IN ANDHERI

At night, it was very quiet and very dark, and you could hear all kinds of night creatures making loud sounds. There were no streetlights or any other building lights anywhere close to our apartments. The sky was dark at night. You could see so many more stars at night and they were so bright. I saw lots of jugnu/fireflies.

Our neighbors were prosperous. They formed a homeowner's association. They arranged private music parties. Babuji loved it. All the kids loved it. But Baiji was not happy. In Matunga, she could shop for everything just by going downstairs. Here it was a long walk on dirt road, and then she had to go all the way to station where all the shops were. She missed her old friends in Matunga. So almost every day, she would travel to Matunga to visit them during day and to do her shopping. Shopping is an everyday activity in India because fresh fruits and vegetables are generally purchased and consumed the same day. She had to be back before it was dark. There was no way she was going to walk in the dark on that dirt road leading to our building. It was so quiet in Andheri. She missed the noise, hustle, and bustle of Matunga. Finally, she had had enough of peace and tranquility of Andheri. We moved back to Matunga.

My parents arranged marriage for my older sister Nirmala/Baby with a Marwari Jain, handsome, tall, very well spoken and very confident young school teacher, Girnar Singh Navlakha, from Indore. Right after her graduation, her Hindu/Jain style marriage took place in Indore on May 9, 1959. They are blessed with two sons, two daughters and a lots of grandchildren and a growing number of great grandchildren.

We still retained the Andheri apartment for about four years after that. My older brother Babu continued to go to Andheri during the day to study. I also went there often. Both of us loved Andheri. We would go there on weekends right after lunch. On weekdays, we went after we finished school. We would come back to Matunga for dinner and to sleep.

GROWING UP IN MUMBAI, INDIA IN 1940S, '50S AND '60S

Babu was the "oldest male" child. That made him special. He was very gentle, very polite, and very good looking. He was fair and had dark, curly hair. He used coconut oil, which made his hair shine. He was Baiji's favorite child, no question about it. He always got new clothes. If certain food items were in short supply, Baiji would make sure to set aside some for him. He was not spoiled by any means, and he never demanded anything of anyone. But he was taken care of by everyone out of love. He loved to doodle in his spare time. These doodles became quite intricate as time passed.

Babu studied for his Intermediate Science Examination at Andheri. I studied for my SSCE/secondary school certificate examination, and for my Intermediate Science Examination at Andheri.

There are many types of accidents which have been almost eliminated in the West and developed countries. However, they continue to occur in India. India is notorious for train-to-pedestrian accidents, which are unheard of in the West. Dangerous railroad crossings, poor supervision of these crossings, and disregard for rules at such crossings by pedestrians cause so many accidents every year resulting in injuries and deaths. Deaths of workers caused by industrial accidents is also high. Unsafe obstacles on sidewalks and roadways cause many injuries, which generally go unreported.

Electricity is provided to Indian residences with a voltage of 220 volts, which is twice the voltage in the United States. Wiring to each floor was done only intermittently in conduits. So you see electrical wires hanging and sagging in some portions. Such open wires can be easily damaged and can become exposed. Electrical shock from 220 volts is a lot more dangerous than a shock from 110 volts. One day, when Babu was alone in Andheri, he touched a live wire by mistake. He was not able to pull himself apart from the wire quickly enough. While he seemed ok in the beginning, the aftereffects of this shock showed up slowly. The exact damage was never accurately determined, but it probably resulted in some combination of internal bleeding, heart damage, lung damage, low hemoglobin count, and poor circulation. Over the next few years he developed many symptoms. His stamina started decreasing. He stayed in bed a lot. He went through a lot of treatment, and after fighting all the ill effects, he died in 1958, at the age of twenty-one.

In his last hours, Babu asked to be given a shower. He asked for sacred tulsi leaves and put them in his mouth. He asked that a bed be prepared for him on the floor. He lay down. Shortly after that, he stopped breathing. He knew he was departing from us forever. He went peacefully.

On that day, I had competed in a fancy-dress competition on the last day of the Navratri celebration, and had won a second prize for dressing up and performing as an African tribesman. I came home expecting to surprise everyone with the good news of my victory. When I reached home at about 11 p.m., however, everyone was crying. I saw Babu lying covered with white sheet. His eyes were closed. Was he sleeping, I thought? But he never slept on his back, and never covered himself with white sheet. I knew he was no longer with us. I did not cry. I could not cry. I just did not believe that he was gone. How could he? He was fine when I left a few hours before, and now he was no more. No, that cannot be true. That simply cannot be true. I went to sleep. Baiji woke me up at about six the next morning. She asked me to go to a close family friend, Matunga wale Mamasaheb, Mr. Mehra. I was to convey the message that my brother was seriously ill, and that Mamasaheb needed to come to our home as soon as possible. This wrong message was an attempt to break the shocking news gently. Mamasaheb knew what my visit meant, and he initiated some activities related to Babu's funeral.

All the major religions that originated in India—Hindu, Buddhist, Jain, Sikh, and many others, believe in the endless cycle of birth and death, until Nirvana or Moksha. Everyone wished for Babu to achieve

GROWING UP IN MUMBAI, INDIA IN 1940S, '50S AND '60S

Moksha. If not, then go to heaven.

At the funeral, people talked in whispered tones about Babu and his many fine qualities. Everyone was sure that he was a great soul. He had to be enlightened to know of his impending death and accepting his death so peacefully. He had lived a good life. He had suffered, but that suffering probably washed away all his bad karmas.

Babu, was laid on a wooden cremation frame. He was covered with a white sheet, except his face which was visible. Flowers were sprinkled on him. I accompanied him and I helped carry Babu's body to a smashan/crematorium near Shivaji Park. I was the oldest male child now, and therefore I had to light the fire that started the cremation. The next day, I picked up his ashes. A month later, I carried his ashes to a holy place, Panchvati near Nasik. I dispersed his ashes in a river and wished him the best for his continued journey in the next life. His final goodbye to this world was in a peaceful place close to Panchvati, where Lord Rama had spent a few of his fourteen years of banishment from Ayodhya.

CHAPTER **20**

New Apartment in Santacruz

AFTER BABU'S DEATH, Baiji was very distraught. She was crying all the time and blaming Babuji for his death and death of her two children previously at or near birth. After Baiji decided that the Andheri location is not right for her, Babuji had continued the search for a new place and kept possession of the Andheri apartment in the meantime, just in case Baiji changed her mind. Now there was no question. We were not going back to Andheri. The search for a new apartment intensified. Babuji knew that getting away from Matunga, where everything was a constant reminder of Babu's death, was an absolutely urgent need for our family.

He found such a place in a brand-new building called the "Shopping Center". This was a building with about 100 small shops, 8' x 8' on average, on the ground floor accessible to shoppers from all four sides of the building as well as from interior corridors. We had now moved into a prestigious brand-new modern building in a very convenient upscale setting. About 75% of all shopping could be done by going downstairs within the building now. We had groceries, clothing, jewelry, appliances, utensils, shoes—you name it—within our building.

This was a vast improvement from the chawl I had grown up in. There was a second building, built at the same time, by the same developer, behind our building. In that building upper floors were offices and extended stay type small rooms. On the ground level, there was a four-star fancy restaurant and bar called "Surang". We could not afford to eat there, but we did go there and just order potato chips. We enjoyed chips with complementary tomato sauce/ketchup, and complementary iced water. In between the two buildings, there was open space. The restaurant used this space for outdoor dining during the evening.

Our building fronted on Station Road, which was a short two block long "L" shaped road. It started just north of the station on the west side of the station. It ran parallel to the tracks, and then to the right, away from the tracks at the railroad crossing. It ended within a block when it met the main road which was then known as Ghodbander Road. The railroad crossing was meant to allow cars and pedestrians to cross the railroad safely to go to east side of the station. This safety was provided by a remote-controlled cantilever lifting gate that was closed when the train was approaching the crossing. It swung open to a vertical position out of the way after the train had passed. The gate worked fine, but many, if not most pedestrians, hundreds of them, continued to duck under the gate, or around it, and enter the prohibited track area. So, a lot of pedestrian accidents happened, resulting in injury and even death. Indian laws limited the Railway's liability for injury and death in most cases. Any compensation paid to accident victims was a token amount, offered mostly as a gesture, rather than fair compensation for damages. No one could sue the railroads. No one did. The railroad always blamed the victims, just said "sorry", and got away with it.

The shopping center complex was built by a Marwari from Rajasthan, who had immigrated to Bombay, just like many millions before him. He came to Bombay poor with a modest dream to make a living. But he succeeded well beyond his dreams. This gentleman started out selling channa-sing/roasted gram and peanuts at Chowpatty

NEW APARTMENT IN SANTACRUZ

beach, from a basket, which he carried on his neck. Soon, he had earned and saved enough money to do other businesses and build this modern shopping, office, residential complex.

The shopping center was the best building in Santacruz when we moved in. Babuji had to pay a deposit, and the rents were high compared to everything else in the area. When anyone found out that we lived in The Shopping Center, they were impressed.

Our building had one and two-bedroom apartments. We moved into a one-bedroom apartment, but quickly transferred into a larger two-bedroom which had three faces of the apartment on exterior walls with lots of windows. In the chawl, we only had two tiny windows, so we kept both the doors open day and late in night to get more light and wind. Here we kept the single door of entry closed. We did not know our neighbors that well, and we had more things to protect. So, in more than one way, we were happy to be in Santacruz, even though we missed our Matunga friends.

We moved to Santacruz halfway through my Intermediate Science, 2nd year of college. Until then, most of my friends were made without any effort. At Kavarana Mahal, I started as a toddler, so most of my friends were naturally made while we were babies. In addition, I made friends at school. Now in Santacruz, I did not have any friends.

I used to commute to college and Andheri from Santacruz by train. One day I saw Dinesh standing on the platform. I vaguely recalled seeing him somewhere. I approached him and asked him where I might have seen him. We quickly figured out that we had a common link. We both studied at St. Xavier's College. This little commonality was sufficient to start a lifetime of friendship. He also introduced me to all of his friends, and we all became a well-knit circle of friends of about 10 young men who saw each other every day and spent most of the evening hours together.

GROWING UP IN MUMBAI, INDIA IN 1940S, '50S AND '60S

Every day after coming from school or Andheri, I quickly finished dinner and went down to the corner of the building. I would meet my friends there, some of whom were already waiting. We would pick up others on our way. Then six to eight from our group would walk to Juhu Beach and walk on the beach until it was dark. Then we would walk back home, and instead of separating when we reached someone's home, we would keep talking, sometimes for another half an hour, before parting. We covered every subject on earth- politics, college teachers, friends, anything, In between, if a pretty girl went by, the conversation would stop. Everyone's heads and eyes would turn toward the girl. Someone would make a comment. Sometimes these comments were made loudly with the hope that the girl might hear it. Sometimes we sang songs loudly to ourselves with the hope that the girl might hear it. The majority of Bollywood songs are love songs with simple but effective words to convey love. So, we could easily pick a song with perfect words. Generally, the girl did not hear our songs. And if she did, she pretended not to have heard it. When we dispersed and went home, it would be about 9 p.m. There was no TV. Radio sucked most of the time, so this was the best use of our time. We spent our evenings this way, but I made sure I did all my study in the college library, or in Andheri, before coming home.

We were not brought up to be fashion conscious. By and large, I wore white shirts, and white, gray, or black pants during the day and night, and pajamas in the evenings and at night. A total of three or four of each was sufficient since clothes were washed daily. They were dried during the day on stretched ropes, and they were ready the next day.

All clothes were made out of cotton and sewn by a tailor who came home with his Singer brand foot pedal powered sewing machine. He made clothes for all the members of our family. He got paid by the day. He was very efficient and produced a whole bunch of clothes at very reasonable cost. While people think of good fashion, when they think of custom-made clothes, we could not care less. These clothes were tough. They lasted a long time despite the fact that they were subject to brutal beatings with a dhoka/a baseball bat style ram in the washing process. They lasted so long that children outgrew their clothes and these clothes were handed over to younger ones.

CHAPTER 21

Railways

WHEN WE MOVED to Santacruz, the railroad became a part of my life in much bigger way. The Santacruz train station could be seen from one of our windows. If you heard the train coming from a distance, it was possible to run, cross the track at railroad crossing, and run onto the platform and catch the train.

The local trains that served Santacruz were part of the Western Railway System. Local trains served by this line all ended at Churchgate Station which is near the Land's End on the south of Mumbai. I only occasionally used a second local system called Central Railway.

The Western Railway System ended on the north side at Virar Station. The stations going home/north from Churchgate were Marine lines, Charni Road, Grant Road, Bombay Central, Mahalaxmi, Lower Parel, Elphinstone Road, Dadar, Matunga Road, Mahim, Bandra, Khar Road, Santacruz, Vileparle, Ghatkopar, Andheri, Versova, Jogeshwari, Goregaon, Malad, Kandivali, Borivali, Dahisar, Mira Road, Bhayander, Thane, Vasai Road, and finally Virar.

There were two types of trains: A slow train, which stopped at every station, and a fast train, which stopped at about half of the stations. Trains ran very frequently, and there was very little waiting for most of the day. Trains ran on a track gage 5'6" wide and was powered by a 25,000-volt AC through an overhead catenary cable. The train system is approximately 150 years old, and you can see signs of wear and tear everywhere. Several millions of people ride this system every day. The trains run on the surface, which is fairly level throughout.

The Santacruz station official entrance on west side of the track was from Station Road. The station was fenced for some distance adjoining Station Building. Shops had been put up illegally all along the face of Station Building, leaving a small opening for train passengers to enter the station. Once you entered, you could buy tickets on the right or continue past a ticket checker who was mainly concentrating in checking passengers who were leaving to make sure they had not traveled without paying for the ride. If you violated this, he would write a citation and collect a fine; or he might let you go with a smaller bribe. If you could not pay the fine, you could go to jail. He spot-checked those he suspected.

Once you entered the station past the ticket checker, you would climb two flights of stairs, approximately 30 feet in height, walk on the pedestrian bridge about hundred feet to your platform, and then walk down to the platform. This was the legal authorized and safe way to do it. But it took too long, and it was tiring. What I am describing is how people, including me, behaved. Unfortunately, after 70 years of progress, some of these bad habits still survive.

When I went to the station, I cut through the ground between the railroad crossing and the platform rather than take the big climb of stairs and long walk. When I got to the station, I sometimes found that the train was still on the platform but had just started leaving. If that was the case, and if the train speed had not picked up too much, I just started running alongside train to pick up speed. Once I was close to the speed of train, I grabbed the handle outside the train door and jumped on to the train. Coming home, I would do the reverse by getting off the moving train as soon as it was at the platform, and it was still moving but slow enough for me to get off.

This is not safe. However, if you ever do it, let me tell you how to do it right. The important part of the technique when getting off is that you must keep running in the same direction as the train after landing on the platform, and slowly bring yourself to stop. If you do not do so, you will fall flat on your face. Newton's laws of motion apply; your head will still be moving close to the speed of the train, while your feet will have stopped. Since your head is still connected to rest of your body, your head has to move forward and your face has to hit the platform.

In the local trains, there were several platforms. You went to the scheduled platform by checking on a display. Sometimes, though, there were unexpected changes. If I discovered that I was on a wrong platform because the train had been switched to another platform, the proper way was to walk up two flights of stairs, cross the track using the pedestrian bridge and then go down two flights of stairs to the right platform. But if you did that, it would take too long, and you would miss the train. So, I just jumped down about 3 feet to the track, crossed the tracks, and then jumped back up on to the right platform. If someone saw me trying to get up, they would always help by lending a hand. You had to do this very quickly because if you could see the train pulling in, you knew you did not have much time.

Looking back, I wonder how I was doing these stupid dangerous things. Only explanation is that "monkey does as monkey sees". Seeing everyone else doing it gives you courage and stupidity to do the same.

RAILWAYS

Mumbai Suburban Train Station. Please note platforms and pedestrian bridge.

In addition to the local train systems in Bombay, I also used the local bus service which supplemented the local train system. The bus was more convenient and very comfortable. However, due to long wait times to get on a bus and high travel time due to traffic and stops every couple of blocks, every trip took too long. I avoided buses as much as possible.

The railroad is a primary mode of transportation for long distance traveling in India. It is also the most preferred way to travel, whenever it is feasible to do so. Traveling by train is very economical, efficient, fast, and reliable. Everyone travels by train in India, even though trains and train stations are crowded and not clean.

Railroad tracks are designed for the single purpose of allowing trains to carry passengers and freight. However, Indians always find many unauthorized uses. One such use was to use the track as a convenient walking trail. People walked on tracks because it is shorter, convenient, and fast. Adults walked on the ties and the gravel between the ties. For kids, it is more fun to walk on the rails. The railhead is narrow,

GROWING UP IN MUMBAI, INDIA IN 1940S, '50S AND '60S

so when you walk on it, you have to balance your foot which takes some practice. When you heard the train whistle, you kept walking and kept looking in the direction of the whistle, which may be in front of you or behind you. When train was close enough to you, you got out of the way. When the train went by you, the train noise increased to a loud rumble and then decreased. I would sometimes put a soda metal cap on the track. After the train had passed, you got the flattened cap which had become magnetized. It was fun to collect these flattened magnetized caps.

Sometimes, passengers pulled the emergency chain and brought the train to stop just so they could get off or on in between train stations. This was never done in Bombay, but it happened on the long-distance trains. This illegal act was punishable with a fine. But when the ticket checker came to investigate, the person pulling the chain would be long gone.

There was no fencing generally to protect railroad property; but in heavily populated areas and near train stations, railways tried to secure the railroad property with a chain link fence. However, someone would always cut the fence. Due to poor maintenance and infrequent checking, this hole in the fence would not be repaired for a long time and would keep getting larger and serve as a passageway onto the railroad property.

Railroad property was also used as a convenient dumping ground. In addition, many people relieved themselves inside the railroad property. They try to face away from the road, so they ended up facing the passing trains. From the train, if you made an eye contact, their expression did not change. They had been doing this for a while. Another use of the railroad property was as an overnight hotel. Many homeless persons stayed on the railroad property for more than a night.

CHAPTER **22**

Education

INDIA HAS A known history of at least five thousand years. Acquired knowledge was preserved mainly through word of mouth and memory transfer from generation to generation and from teachers to students. The second major source of preservation was religious texts, which have been preserved by periodic recompilation by scholars and religious leaders. Based on such preserved information, it is concluded that for much of this period of several thousand years, and even now, most Indians divide themselves into castes, even though discrimination based on caste is illegal per the Indian Constitution. Details and severity of discrimination has fluctuated over this history. For about the last thousand years, four castes have been recognized. Even though after the independence in India, attitudes have changed, the caste system still survives, in practice, more than seventy years after independence. While the caste system is a Hindu concept, many Indians with non-Hindu faiths such as Jains, Sikhs, Muslims, Jews, and Christians, also practice it in some manner. The level of education and type of education a child might receive has been, and continues to be, influenced by the caste the child is born into.

The cast of Brahmins, which is considered the highest, got the best education in religion, art, and science, and this caste in turn provided most teachers in religion, art, and science. Brahmins exclusively performed public religious services. Brahmins were not paid very highly, but they were highly respected. They often performed these services out of sense of duty, for the pleasure that came from their work, and also for respect and admiration they got for their knowledge.

The second highest cast was that of Kshatriya. This is the caste that maintained law and order, defended the country, and waged war. They got special education related to these duties. This was the most powerful caste in terms of power. The king, most of the king's advisors, and most of the soldiers belonged to this caste.

The third highest caste was Vaishya. This is the business class that did all the manufacturing and trading. This is the class that was generally the most prosperous. I belonged to this caste. This caste got some basic education such as reading, math, etc., but practical experience gained while working in the family profession was emphasized the most.

The fourth caste was Shudra, or labor class. This is the class that was poorest, least educated, and always oppressed. They were discriminated. This class used to be referred to as the untouchables because many from other three castes believed that touching one of them made you so impure that you were supposed to take a bath afterward to cleanse yourself. Many from this fourth class converted to other religions, including Christianity and Islam, to escape this discrimination. During the six hundred years of Islamic and Christian rule in much of India, those who converted to Islam and Christianity received favorable treatment from the government and thus were able to improve their lot.

After the independence, discrimination against the Shudra was banned. Many affirmative action programs were introduced by the government to correct past discrimination. A favored class was created by new classifications such as "Backward class", "Schedules

EDUCATION

class and tribes", "Other backward classes", etc. These efforts have worked. The Shudra caste has progressed economically and is fairly strong politically. So much so, that many people from three higher castes, particularly from the Brahmin Class, protest that these affirmative actions have been overdone and they are a form of reverse discrimination. Some, belonging to one of these higher castes, have also tried and succeeded in reclassifying themselves for favored special treatment in jobs, school admissions, etc. New groups continue to appear and appeal to the government to reclassify them as scheduled or other backward classes. What started out as a necessary and reasonable program to help the minority has now ballooned into a special favored class. A student belonging to this class gets favored admission into schools or colleges and scholarships. Others get favored treatment in government jobs and many social programs.

My father was well educated by the standard of the Indian society at that time. He had a good command of Hindi, English, and mathematics. He had a good job, and he was going to educate his children as much as they wanted. I did not need any special treatment to get into a school. And I did not need any financial help from government.

As I mentioned earlier, I got my education started without any pre-kindergarten or kindergarten. In those days, kindergarten was available only in a very select elite neighborhood. In the chawl that I grew up, no one could afford it. So, I started my education with first grade. I started about a year and half younger than I qualified, and I advanced to third grade in the middle of my second grade by taking an exam and passing it. In the seventh grade, I convinced Babuji to put me into a private English language class conducted by an Arya Samaj Group in the next building, which was named Rajmahal, or king's palace. I took these classes as additional classes. There was no school credit for these classes. My teacher was excellent. By the end of the course, I could talk and write well, and even started thinking in English when I was alone.

GROWING UP IN MUMBAI, INDIA IN 1940S, '50S AND '60S

This early proficiency in English helped me a lot throughout my life. For example, Hindi had been selected as a national language, and there was a great effort by the government to push everyone to learn Hindi as soon as possible. I was studying in a Hindi medium school. Even so, I could not find good physics, chemistry, and astronomy books in Hindi, when I studied for my high school matriculation exam. I studied these subjects in English and took the exams in Hindi. I studied arithmetic in Gujarati. I read additional books for science in Marathi.

I took my final high school examination called Secondary School Certification Examination, or SSCE in the 11th grade in March of 1958. Until the 10th grade, all the exams were set by, conducted by, and graded by my schoolteachers and held in my school. But in SSCE, all the exam settings and grading were done by teachers under independent contracts and under the supervision of the SSCE board at locations away from your school as designated by the SSCE board. The results were announced in a newspaper using a designated registration number. Results were published in the morning newspaper of a large circulation paper like *Times of India*. The registration numbers of all passing students were printed, but the registration numbers of failing students did not show up. This was done roughly in the beginning of May. After about fifteen days, you could pick up your actual certificate and mark sheet from your school. Mark sheets in addition to actual marks also noted whether you placed in first class with distinction, first class, second class, or pass class. I had done well, and I was happy. I passed in first class with distinction, which required that you obtain 70% or more marks/points. This put me in a select group of the top 3% to 5% of all the students taking this examination in the entire state.

Almost everyone who passed these final school exams continued their education into college. There were three primary choices - Arts, Science, and Commerce. Those who took Arts generally made very

EDUCATION

little money in their career. Being from a vaishya/business caste and also having a dad who was employed in business, I ruled out Arts. Commerce might have been appropriate for vaishya/business class, but it seemed boring. I could not bear the thought of working repeatedly with monotonous numbers all day long. When I thought of Commerce as a career, my memory of the tedious and boring times when I helped Babuji with the general ledger for his business came back. So, I ruled Commerce out. What was left was Science. Going to a science college was an obvious and easy choice, and my final choice.

Since I was in top 3%-5% and passed the SSCE with distinction, I could choose any college in Bombay and be assured of admission. Elphinstone College, a government college, was considered the best and attracted the best students. I chose to skip it because of its reputation as a dull place. I next considered the second best, a Catholic Church sponsored college named St. Xavier's College. It had a good reputation for scholastic as well as non-scholastic activities. Many of the professors were European missionaries, and it had a good reputation. I heard that there were lot more girls at St. Xavier's compared to Elphinstone, Also, Elphinstone had a reputation that only nerds went there. This helped me decide on St. Xavier's College. When I applied for admission, the college required a parent's approval evidenced by a parent's signature. My dad was not with me. There was no time to wait until the next day, so I forged his signature, and I was admitted. What a way to begin! I was in a good college, I was a good student, and I was hardworking. I started college with great optimism that I had a good future ahead of me.

Being the oldest son, I assumed responsibility to ensure a good education for my brothers and sisters. This was important for their good future. Unemployment in India, particularly among the young, was very high. Many Bollywood movies were centered around this subject. They showed how unemployment led to drinking, crime, misery, and even suicide among youths. I was determined to keep that from happening to my siblings. So, I spent time helping them learn. But I also felt that I had to be a disciplinarian. I tried to balance that by offering rewards out of my pocket money in a kind of carrot and stick approach.

CHAPTER **23**

Girls

BY THE TIME I finished high school, I had spent 5 years of my young adulthood in an all-boys school. So, there was no interaction with girls at school. When I started at St. Xavier's college, I was really looking forward to making up for those five years of all-boys school. I was going to make friends with girls.

Let me tell you, I was not the only one with this plan. However, there was one problem. In those days, and even now, though to a lesser extent, girls are raised in India in such a manner that when the time comes, they will get the best match in a marriage arranged by parents. This means that their reputation must remain pure. Any known interaction with boys, even verbal, may be viewed with suspicion by the prospective bridegroom or his family.

Therefore, girls were taught to just ignore boys and walk away. Fear of rejection or being ignored kept boys from even trying. This lack of confidence was often covered up as shyness or lack of interest. But the fact was, it was nothing, but lack of guts, to face rejection. Well I had the guts. So, I did make friends with girls. At least they talked to me. They were not girlfriends by any means. But talking to them as my friends brought me a lot of joy.

GROWING UP IN MUMBAI, INDIA IN 1940S, '50S AND '60S

My friends and I would often share stories of our "adventures" with girls. My friend Kishor told us his story.

One day, Kishor was on the third floor standing against a railing looking down into the central courtyard of his college. He saw a girl on the second floor below, about a hundred feet away. He looked at her, and she looked back at him. Their eyes met for a second. As soon as that happened, she turned her eyes away; she turned and then walked away. I was waiting for him to continue with his story. But he was very quiet. I asked him why he was quiet, and why he was not telling us what happened after that? He said that he would never forget that moment when their eyes met. I needed to know more. So, I prodded him. I said tell me more. Tell me about the time you met her again. He said he never met her again! A second's worth of meeting by eye made memory of a lifetime for him. You figure it out! Only in India of 1962!! Just knowing that a girl may be aware of your existence was enough to make you fall in love with the girl: Even if you never met her after that; even if she never knew how you felt.

Many girls probably were also experiencing similar love. My friends would often claim that they met a particular girl somewhere. When I asked them for details, like what did you talk about, they would clarify that what they meant was that they saw this girl from a distance. The fact that the girl was not even aware of their presence did not seem to matter. These guys were so desperate to mislead and exaggerate. What can I say?

We were taught to think of all girls as our sisters, particularly the girls in our neighborhood. We were also expected to look at the girls from neck and above, as if the rest of the body did not exist. Regardless, girls and women were interested in looking beautiful and attractive. Girls did not wear much make-up, because parents discouraged it. Use of face powder and lightening cream was allowed and it was common. Use of lipstick was not. After marriage, many husbands did encourage their wives to wear make-up. Use of single flowers in hair was also very common. But use of flower garland called Gajra was much more common. Jasmine or Mogra was the flower of choice. It was very common for husbands to pick up a Gajra from a street vendor, on their way home after work, in hope of a happier wife and an improved night.

CHAPTER **24**

Science College - St. Xavier's College

AFTER I FINISHED high school and the summer vacation following the school was over, and after I was admitted to St. Xavier's college in science studies, I was going to begin a new chapter in my life. College can be tough compared to high school for many reasons. The first difference was that the medium of education was now English. Most students studied in Indian languages in high school. I studied in the Hindi medium at Marwari Vidyalaya High School. However, the extra private classes I took for English at Raj Mahal when I was in seventh and eighth grade laid a good foundation for me and improved my ability to read, speak, and write English. So, change of language after high school for almost everyone turned out to offer me a competitive advantage.

My continued interest in acquiring a good command of English by reading English newspapers and magazines and using the language whenever there was an opportunity, I knew was going to help me in college and the rest of my life.

Another difference between high school and college was that college classes were huge. In many subjects, there were two hundred stu-

SCIENCE COLLEGE - ST. XAVIER'S COLLEGE

dents in one class, compared to about twenty-five at my high school. If you got the seat in the back, sometimes it was hard to see, hear or understand the professor. When it was hard for me to see what the professor was scribbling on black board, I thought it was normal not to be able to see everything clearly from back of the room. I had never worn glasses and had not seen many children wearing glasses. I did not realize that I had become near-sighted. For almost two years in college, I managed without glasses, with bad vision. Even so, I did very well. In algebra I stood first in the whole college. Generally, I was near the top of my class, overall.

At the end of the first year of science college, a decision whether to pursue engineering or medicine had to be made. Depending on your decision, you would drop biology or mathematics. In biology during the first year in science, we had to dissect a frog. I am a Jain and Jains never intentionally hurt any living being. Being vegetarian and being Jain and because I never had any pets, the dissection of animals, birds, or even small creatures, such as a frog, was a very unpleasant experience. While I did not throw up, as some of my classmates did, I decided that medicine was not for me. I was excellent in mathematics, so my decision to become an engineer was easy. I dropped biology.

St. Xavier's college was founded in 1869, in close collaboration with the Society of Jesus in Germany and Spain. We had number of professors who had title of Reverend. They wore the white robe of a Jesuit minister. We also had non-Christian Indian professors. Most professors were great. One of them was Professor Kothare who taught Inorganic Chemistry. He had a very good command of English. He had studied in England and he delivered a fiery lecture on Inorganic Chemistry in a powerful and almost poetic English with a sing song delivery. His vocabulary was exhaustive. He spoke with great fluctuation of tone and pitch. He had spectacular gestures. And he did that while he was explaining periodic tables! I cannot describe how someone can make a subject like Inorganic Chemistry so interesting. He kept us spellbound. I will never forget him. On the other hand, there were some professors who could make you doze off. But not Professor Kothare. He would keep you on your toes throughout his lecture. When he finished his lecture, you were exhausted and exhilarated at the same time.

I loved the stone architecture of St. Xavier's College, which, for most part, had been built between 1869 and 1937. It was often used by Bollywood moviemakers as an outdoor set. In 1959 and 1960, as I passed through the open courtyard to go to my classes, or where I would meet my friends, I had no idea that exactly fifty years later, the first black President of the United States, President Obama, would be meeting students in Bombay in a townhall meeting at that very spot. A year later Hillary Clinton would do the same as the Secretary of State of the United States.

Barack Obama greeting students at townhall meeting at my St. Xavier's College in 2010.

SCIENCE COLLEGE - ST. XAVIER'S COLLEGE

My college had a very nice large library where I studied almost every day; in between classes, and at the end of the day. I never left until I had done everything that I wanted to accomplish. The library was well stocked with a wide variety of books. This library created the hunger for learning which has lasted for a lifetime. The joy of knowing something I did not know before, would rise inside me like a tidal wave. I have not experienced a greater joy than that. When I learned something new, it made me feel very happy and very special. It still does. I was very proud of my college. It had everything, including one of the few basketballs courts in Bombay!

Even though St. Xavier's college was founded and run by Jesuit priests, religion was not discussed in classrooms, with one exception—we had one hour per week of compulsory religion class. In this class conducted by a white reverend, there was an attempt to expose students to Christian thoughts. I had a similar class at Marwari Vidyalaya, where a Hindu teacher would explain Bhagwat Geeta. So, I was delighted to learn more about another religion. In one of the classes, I asked the reverend a question, "how is it that killing animals for food was allowed and even encouraged by your Christian religion that valued kindness very highly". His answer, while quoting from the Bible, was that everything, including all the creatures, were created by God for human benefit. The proof according to his explanation lay in the Christian belief that God created man on the sixth day of creation, after He had created the rest of the earth including all of its creatures, roughly six thousand years ago. I thought this argument was very selfish and self-serving and the timeline of creation so inaccurate. But I did not argue my point any further.

Professor Agrawal ran a private mathematics coaching class for the intermediate final exams. Results of this exam determined which college of engineering would accept you, if any. He had a very high success rate. So, in the summer vacation following my first year of science, I decided that I wanted to take his classes. Unfortunately, both

GROWING UP IN MUMBAI, INDIA IN 1940S, '50S AND '60S

Baiji and Babuji had gone to Madhya Pradesh for a wedding. Baby, my older sister, was in charge at home. I needed two hundred rupees to register and she did not have it. And there was no time to arrange for it. However, my younger brother Sushil had saved enough money from his allowance to temporarily finance my education. Amazing that my younger brother, a high school student, could finance my education.

Professor Agrawal taught me mathematics. But more than just mathematics, he taught me how to succeed. His method was very simple—think smart and work hard. He gave you a lot of homework. You were expected to do all the homework in the evening after classes. The following day, he would solve some problems very quickly on the board; this student had a problem with. If you had done the homework, and if you did not understand some part of the solution, you could ask, and he would explain. But if you had not done your homework, you would be removed from the class. He had no patience, and he was not going to waste his time on losers. He would give every possible variation of all possible questions to you. Once you did that, you really had mastery over the subject. From the long list of students wanting to take his classes, he chose only the brightest. And then he made these smart kids work hard. No wonder, most of his students performed very well in the final exam.

There were two things that I learned from Professor Agrawal. One is that you can learn anything you want by yourself. That is what he made you do by giving you homework before he had taught you the chapter. You don't have to have a teacher. A good book will do. Second was that you can learn when you want to learn. You do not need to restrict yourself to a designated time, day, year, or time in your life.

In the second year of college we had a lot of laboratory practical classes. Spilling chemicals on clothes and getting holes in them was pretty common. In physics, I was having some difficulty looking through optical instruments. This was puzzling and disappointing because I did well in everything else. The answer came one day in an unexpected way. I spotted a friend, a girl, coming from a distance. I was so happy to see her that I started waving at her furiously as soon as I spotted her. When she got closer, I realized that I was waving at someone I did not know. I felt stupid. She looked at me, rolled her eyes and walked by. I knew I needed glasses, and an eye examination confirmed that. I got my glasses and have been wearing glasses ever since.

CHAPTER 25

Engineering College - VJTI College

BEFORE INDEPENDENCE, MOST of India's population was involved in subsistence level small farms. A small percentage was in light manufacturing, trading, and government services. The elites either worked for the British government or were landlords, or businessmen with good connections with the British government.

All this was going to change with the independence on August 15, 1947. There was great optimism that there would be opportunity for everyone. And everyone knew that to succeed, you would need a good education. Twenty-one years after the independence, in 1958, when I was entering college, India had progressed. But there was great unemployment among youth. If you did not get a college education, or if you got college degree in arts, your chances of finding employment was very low. Those who got a science or commerce degree did much better. Those who got a medical or engineering degree did the best.

Based on the above situation and other considerations, I had already decided that I was going to be an engineer. But I was not sure what kind. I asked senior people around me for advice. But their advice

ENGINEERING COLLEGE - VJTI COLLEGE

was confusing, biased, and often depressing. It seemed that most people who I talked with were unhappy with their choice of career. So, they tended to steer me away from their career.

I decided to do a little research on my own. I went to the Vocational Guidance Bureau, which was located inside the Central Unemployment Office in downtown Bombay. The unemployment office also had a small library, which had some books on vocational choices. I read some of these books. In one book, most of the engineering disciplines were listed, along with description of the kinds of work engineers performed, and the range of expected income levels. Based on this additional information, I decided to become a civil engineer.

Only a small percentage of those who were lucky enough to get into science college applied. Of those who applied only a few got into engineering college. There was only one engineering college in the largest city in India with a population of 10 million at that time. So competition to get into engineering college was fierce. The only criteria of admission were your grades/marks. There were no interviews or any other evaluation methods. Everyone tried hard to get the best marks in Intermediate Science. I did the same. My hard work studying day and night, both in Matunga and Andheri, and taking Agrawal classes, paid off. I finished near the top. I entered engineering college, named Victoria Jubilee Technical Institute, VJTI, as the third best student of my University of Bombay Civil Engineering freshmen class of about thirty students.

VJTI, which was later renamed Veermata Jijabai Technical Institute in 1998, was founded in 1887 with 2 departments—mechanical engineering and textile engineering and had subsequently expanded to include civil engineering and electrical engineering.

GROWING UP IN MUMBAI, INDIA IN 1940S, '50S AND '60S

VJTI, Engineering College, Veermata Jijabai Technological Institute, named Victoria Jubilee Technical Institute when Author attended it.

There were examinations throughout the year. These exams were mainly to help students' study throughout the year, and also to help professors assess how students were doing and to help them improve, as necessary. However, the only examination that counted, for pass/fail and grades, was the final one at the end of each year. This was the system throughout my schooling from primary school to engineering college. As a result of this system, many students did not study hard until the final exams were very near. It was very common for students to skip class, even when they were present in school. It was considered "cool" to skip classes.

Even though I was much more serious than most about getting a good education, and I was interested in learning for knowledge and not just to pass the exams, I had my share of skipped classes. I wanted to be cool like my peers. If the professor was boring, or not very knowledgeable, students were more likely to skip. Just by estimating the attendance, you could estimate how good the professor was. Professors tried to devise methods to increase attendance, which included punishment and other methods, with limited success.

ENGINEERING COLLEGE - VJTI COLLEGE

During the first year in engineering, I learned at the pace the professors set. After they taught something, I would then try to understand it better by reading the material again. After I understood, it was important to memorize it so that I could recall it at the end of the year in final exams. In order to facilitate recall, I prepared the material in re-written short notes/points. I would refresh my memory by looking at my short notes just before the exams.

In the second year of engineering, I improved my study method. After I completed the first year of engineering, I realized that I needed to do what Professor Agrawal had taught me. I bought or borrowed textbooks ahead of time and completed quite a few subjects on my own during the summer vacation before college started for second year and third years in engineering. This method worked very well. I prepared myself ahead of the classes by studying many subjects during the previous summer vacation.

I kept up with my studies by staying in college most of the day, spending a lot of time in library, in between classes and after classes. Only after I had completed all my studies did I go home in the evening to eat supper, to relax, and to sleep. I asked smarter questions, and gave smarter answers in class, impressing everyone. Unlike most of my friends, who were very tense as the final examinations approached, I was not. I was well prepared. Because of this, even though I had pneumonia just before the final exam in the second-year engineering, I was ranked #1. And my marks broke many records for our college.

In one class, Hydraulics or Fluid Dynamics, I knew the subject well enough to point out professor's mistakes so many times that my classmates gave me a nickname "(Professor) Levitt". Professor Levitt was one of the authors of our American textbook on hydraulics. These books were made affordable by publishers charging a lower price for a paper back/soft cover "Asian" edition printed and sold in India, and some Asian countries.

In another class, Electrical Engineering, my professor asked a very difficult question which was to be followed by an explanation. His expectation was that no one would answer, and then he would explain. Before he had completed his question, I blurted out the answer. He was very surprised, almost shocked. He was very happy. He started praising me effusively. He used superlatives like fantastic, unbelievable, incredible, etc., to describe my answer. He went on so long that I got scared. I thought that he was setting me up for ridicule later. But no, he was truly praising me. I think it is rare to find a teacher who can praise a student so enthusiastically. I will never forget him and that was a proud moment for me.

I was not good at everything. One of the courses I was required to take was called workshop/practical engineering. I was an engineering student in civil engineering, But I had never driven a nail or turned a screw. It was not necessarily my fault. It was just the way things were. Labor being cheap, people never learned to do things for themselves. A carpenter would be called every time some work needed to be done, no matter how simple. Indians generally were not handy, and most households do not have a lot of tools. I only saw a couple of screwdrivers, a hammer, and a plier. A few times that Babuji had attempted some projects, they ended up disastrously. So, when I got a chance to do workshop, I was quite excited. We were required to wear a prescribed dark blue workshop cover-all suit which protected your clothes.

By rotation, we had to do projects in carpentry, foundry, and machining. Projects were to be done by a team of two, which was formed based on an alphabetical list arranged by last name. Khan preceded my last name Kothari, so I partnered with Khan. Khan was tall, strong, and handsome. I was tall by Indian standard at 5' 9". But he was really tall at 6' 1". I was slightly above average Indian at 115 pounds, but he was huge at 150 lbs. We made a good team. He had a good sense of humor, he cared for me a lot, and he did most of the work. I mostly watched.

ENGINEERING COLLEGE - VJTI COLLEGE

In foundry, our project was to bend a 1/4-inch diameter rod into a shape of number 8. The process was to heat the rod to a red-hot color in a coal burning furnace. You grabbed the red-hot rod with pliers and set it on an anvil. One person held it in position, while the other hit it with a sledgehammer to gradually bend it into a circular shape. You repeated the process of heating and hitting until you had your number eight. Then you dropped it in water to cool it and submit it to your teacher. While we were supposed to alternate the tasks so that both of us would learn, Khan did most of the hitting. When I hit with the sledgehammer, the rod did not bend at all!

In 1960, I discovered a rare baby picture of me. For fun, I took it to my college and shared it with Khan. Khan asked me if he could keep it for a day or so, and I agreed. He forgot to return it- I forgot to ask for it, and I never got the picture back.

I left India in 1964 and started my own consulting business in 1976 in Unitd States. During first few years of my new business, I did extensive traveling to promote my business. I and Khan were not in touch after I left India in 1964. I came to know that Khan had also arrived in USA and was living in Queens in New York City. In one of my business trips I decided to visit Khan.

When I arrived at Khan's home, after we greeted each other with great enthusiasm, he asked me to stop. He said he had something of mine that he had meant to return for a long time. He handed me a picture. I was amused and happy to see my baby picture that I left mroe than twelve years arlier.

I had no idea that more than more than 40 years later, I would use this same picture on front cover of this book.

Many kids used to have their homemade lunch delivered by a tiffin wallahs/courier. I was one of them.

GROWING UP IN MUMBAI, INDIA IN 1940S, '50S AND '60S

Tiffinwallahs are unique to Mumbai. Thru a network of Tiffinwallah union, they deliver your hot lunch from anywhere in Mumbai to anywhere in Mumbai within 3 hours by 12PM at a vey nominal cost. This is accomplished by your "Tiffin' going thru 5 or 6 transfers at various connecting node points.

So after eating a homemade heavy lunch, it was hard to stay awake in class, particularly if the professor was boring. Once one student yawned, more kids would start yawning. Yawning is so contagious. It was funny to see the helpless professor trying to control this hilarious situation.

CHAPTER 26

Trip with Friends

DURING SECOND YEAR of engineering in 1962, we had a short vacation coming up coinciding with the Diwali festivals. My friend Kanti proposed that we should take a trip to see the north part of India. A few friends agreed. I was one of them. We ended up with a group of five. Kanti was the planner and our leader. Ashish, Ajitabh, Dipak, all Gujarati, and I who spoke good Gujarati and was often mistaken for Gujarati. We formed a very good compatible group. Throughout the trip, we were very accommodative of each other. Amazingly, we did not have a single fight, even though we spent seventeen days traveling—exhausting traveling, sometimes without a proper meal, and often with all the hassles that go with travel. My friend Kanti loved to sing Mukesh songs. I also liked to sing, and this made our trip even more enjoyable.

Kanti planned everything to perfection. He made all the reservations. He prepared a very detailed packing list breaking it down for each person's share. The list was perfect. His hard work made our travel smooth and very enjoyable. We appreciated his hard work.

We were born in India at a time when there was great unemployment and wages were low. We felt an obligation to minimize the financial burden on our parents. We were brought up to be careful with our money. We pinched every paisa/penny when we could.

GROWING UP IN MUMBAI, INDIA IN 1940S, '50S AND '60S

During the British period in the late 19th and early 20th century, the term coolie referred to an indentured Indian laborer in South Africa and other British colonies. Indenture laborers were contract laborers. The contracts were one sided, in favor of employers. Contracts were allowed to be sold or transferred. Coolies were treated almost like slaves.

However, coolies are now honorable professionals of porters at railway stations, bus stations, and airports. These porters carry your luggage from or to your taxi, rickshaw, bus, or train. The profession is mostly unionized in India now. But when we took this trip, it was not. Due to high unemployment and easy availability of cheap labor, everyone except the very poor used coolies to carry their luggage. Anything that weighed more than fifteen pounds was mostly carried by a coolie.

When we had to hire a coolie, we negotiated tough. We generally offered to pay him less than half of what he was asking. We also hired coolies to run into the arriving train and claim some seats for us if we did not already have reserved seats. Once we occupied these seats, the coolies brought our luggage and stacked them onto the luggage rack. We paid them after the service was complete.

Kanti, in consultation with the rest of us, planned everything meticulously. He studied the railway routes and made the most efficient travel plan. The Indian railways had a highly subsidized flexible travel plan for students. We took full advantage of it. He made all the reservations, and if our plans changed, he made the appropriate adjustments. All the tickets were transferrable to a different schedule, if available, generally at minimal additional charge.

Everyone carried two pieces of luggage. Clothes and miscellaneous items were all packed in a painted steel trunk. The second piece was our rolled bed which consisted of a thin fluffed cotton filled mattress, a pillow, and a blanket. A pillow was optional. Some of us used our shoes or clothes as a pillow substitute. A bedding with thicker mattresses and sections for storage, rolled and secured with leather straps was used by our parents, but they were too bulky and too expensive for us.

We stayed in free or cheap hotels as much as we could. Often, we stayed in a dharmshala, a government Dak (post) bungalow and military camp. These are accommodations that charge a very nominal rent. A dharmashala is built with donations from rich generous individuals and are run as charity and to earn "punya/good deed" points for community service. According to Hindu tradition, helping a traveler by providing him with food, water, or shelter is considered a very high form of punya—good karma/deed. A person who accumulates a lot of punya is more likely to go to heaven.

We ate in reasonably priced restaurants. Our fun in this trip was not going to come from the comfort of travel or the quality of the food we ate. It was going to come from visiting the exciting places we were going to see, and from different types of people we were going to meet.

Our trip was primarily for enjoyment as tourists, but it was also for education. Since we were all classmates studying civil engineering, we included the United States Embassy in Delhi which was constructed in 1954. We had read rave reviews about its architecture. It was designed by a famous American architect, Mr. Edward Durell Stone. This was one of the projects that he was particularly proud of. He went on to design many famous projects during his personal consulting practice from 1932 until his death in 1978. Another project that we included was designed by a French Architect Le Carbusier. We also included the Howrah Bridge in Calcutta.

Our itinerary, which started in Bombay and returned to Bombay, covered following locations in the order of travel: Lucknow, Delhi, Agra, Fatehpur-Sikri, ChandiGarh, Deharadun/Rhishikesh/NainiTaal, Allahabad, Benaras/Varanasi, Calcutta, Shanti Niketan, Darjeeling, and Gangtok/Sikkim.

Lucknow:

Lucknow is renowned for Urdu literature. I read Hindi novels, Hindi poetry, and watched Bollywood Hindi movies. Hindi and Urdu are very similar. Hindi is written in Sanskrit or Devnagari script, Urdu is written in Arabic script. Hindi is heavier in use of Sanskrit words; Urdu is heavier in Arabic and Persian words. A Hindu will generally claim that he is speaking Hindi and a Muslim will generally claim that he is speaking Urdu, even though in everyday conversation, you notice little difference.

Lucknow is the capital of Uttar Pradesh. One legend has it that Lord Rama's brother Laxman established the city. The city's name evolved into Lakhanpuri which the British changed to Lucknow. Lucknow is a major metropolitan city of India and may be in the top ten most populous cities of India. It is the most populous city in Uttar Pradesh, which is also the most populous state of India. Lucknow has always been known as a multicultural city; it flourished as a cultural and artistic hub of North India and as a seat of Nawab power in the 18th and 19th centuries.

It is a common perception that everyone in Lucknow, from Nawabs down to ordinary person, is very polite and very considerate of everyone else. The Nawab title is a Persian/Arabic word originally given to rulers of princely states by Mughal Emperor, and later adopted by British rulers for similar purpose. A common joke connected with Lucknow is that two nawabs arrive to board a train and are at the door to board exactly at the same time. Out of courtesy, one of them gestures an invitation to the other person to board first. The second person responds to the courtesy by insisting that the first person board first. The first person, in turn, responds by insisting that the second person board first. While these two gentlemen are taking turns in being polite and considerate and fighting to be more courteous than the other person, they remain unaware that the train had already left.

Lucknow was our first stop. We were traveling on a very tight budget. Saving money whenever we could had become our habit. Our first lunch was one item: Yogurt, period. We saw various parts of city, including Imambada.

Delhi, Agra, Fatehpur Sikri, and Taj Mahal:

Through most of history of more than two thousand years, Delhi and Agra, a region with about 100 miles distance between them, has often served as a capital of various kingdoms and empires. Control of the Delhi/Agra region has passed through many different rulers. Most of these rulers destroyed, rebuilt, or added to various parts of the region. You see their influence everywhere.

It is believed that Indraprastha, which was built by Pandavas of Mahabharat epic, was built in this region after they cleared existing forests by burning them. There is archeological evidence of the Maurya period which produced the great King Ashoka around 300 B.C. King Anang Pal of Tomara Dynasty founded the city of Lal Kot in this region around 700 A.D. The city was conquered by Prithvi Raj Chauhan around 1200 A.D. and the name changed to Qila Rai Pithora.

The Defeat of Prithvi Raj Chauhan started a long period—approximately five hundred years—of Muslim Dynasty rulers from the Afghanistan, Uzbekistan, Turkic, and Mongolian regions. Muhammad Ghori from Afghanistan, followed by his General, Qutbuddin, started this long Muslim rule. Qutbuddin built Qutb Minar. Razia Sultan, his granddaughter, was the first and only Muslim woman ruler over Delhi. Jalaluddin Khilji, Alauddin Khilji and Muhammad Bin Tughluq followed. Muhammad Bin Tugluq moved the capital of his empire to Daulatabad in Maharashtra. This was a bad move. He lost a lot of his northern territory which forced him to return to Delhi. But then he lost much of his southern territory. Taimur Lang (Timor), who committed a great genocide of tens of thousands of captives was followed by the return of the Lodhi Dynasty.

The Lodhi Dynasty was followed by the Mughal Dynasty which ruled for a long period of almost three hundred years. Babur, founder of the Mughal Dynasty, was the descendant of the Mongolian Invader Genghis Khan from Mongolia and Taimur Lang from Uzbekistan. Babur was successively followed by Humayun, Akbar, Jehangir, Shahjahan, and Aurungzeb. Shahjahan built a new capital known as Shahjahanabad, now known as Old Delhi. After the death of Aurangzeb, the Mughal Empire became weak and the Maratha Empire had major control over a great area including Delhi for a while. Nader Shah of Persia looted Delhi and carried away a highly-jeweled peacock throne and one of the biggest diamond Koh-i-noor which ultimately ended up under British control.

The British found a way to control India when Hindu and Muslim rulers of India were weakened by lack of unity and infighting. The British ruled India for approximately two hundred years until August 15,1947 when independence was granted and the country was split up into New India, Pakistan, Ceylon, now renamed Sri Lanka, and Burma, now renamed Myanmar. The partition of India and Pakistan caused hundreds of thousands of Hindus and Sikhs to migrate from Pakistan into India. Many of the refugees from Pakistan ended up in the Delhi region. This migration was followed by an economic migration from various parts of India. As a result, the population of Delhi Metro region has surged to approximately twenty-five million people today.

The iron pillar of Delhi, constructed in 400 A.D. by Chandragupta

TRIP WITH FRIENDS

Vikramaditya Maurya of the Gupta Empire, is one of few structures of the Hindu period. Most of what we saw in our trip was built by Muslim or British rulers. We saw the India Gate and the Parliament complex built by the British. We saw the Humayun Tomb, Qutub Minar, and the Red Fort built by Muslim Rulers. And we saw the American Embassy. We walked in Connaught place and we ate in Chandni Chawk.

When we were waiting for a bus, I was impressed by the fact that people were waiting for their turn and not pushing to get in first, like they did in Bombay. Remember this was in 1962. Delhi was much smaller than Bombay then. I talked to a college girl while both of us were waiting for a bus. Her Hindi was so pure and pleasant. Hindi in Bombay was corrupted by words from so many languages spoken by cosmopolitan Bombayites. Marathi had the greatest influence.

The Moghuls built Agra Kila. Agra was the capital of Mughal Empire under Akbar, Jahangir, and Shahjahan, and was known as Akbarabad. Babar laid out the first formal Persian garden on the banks of river Yamuna, called Aram Bagh. Akbar made Agra a center for learning, arts, commerce, and religion. Akbar built a new capital city on the outskirts of Agra called Fatehpur Sikri. However, he abandoned it after a very short time and returned to Akbarabad. Jahangir had a love of gardens and flora and fauna and developed many gardens inside the fort.

Shahjahan, known for his keen interest in architecture, gave Akbarabad its most prized monument—the Taj Mahal. Built in loving memory of his wife Mumtaz, the mausoleum was completed in 1653. Shahjahan later shifted his capital to Shahjahanbad in Delhi. His son Aurangzeb moved the capital back to Akbarabad, usurping his father and imprisoning him. Akbarabad remained the capital of India during the rule of Aurangzeb until he shifted it to Aurangabad. Agra is the birthplace of the religion known as Din-e-Ilahi which Akbar promoted. Din-e-Ilahi tried to combine the best of Hinduism and Islam.

GROWING UP IN MUMBAI, INDIA IN 1940S, '50S AND '60S

We did sightseeing in Delhi by bus tours and local buses. We went to the Parliament complex, Gandhi Memorial, Connaught place, and Chandni Chowk. We took additional tours to see the Taj Mahal, Agra Kila, and Fatehpur Sikri. Seeing all these great monuments made me wish for the past. But I also believed that we were going to achieve a lot as a democracy for all of the people of India, rather than just a few who were privileged to be in power.

Taj Mahal, Agra, India

Chandigarh:

As a budding engineer, I studied architecture as well as engineering. I was particularly looking forward to a French architect, Le Carbusier's, design for many buildings which we had read about. On seeing the State Legislature building from the inside, I was disappointed that the Indian government and the Punjab government actually carried out Le Carbusier's design. I had learned that form must follow function. To build such a large and tall empty space, with the inherent extra cost of construction, cost of cooling the space, maintenance of all the tall concrete walls that would crack soon in the hot Chandigarh sun, was not justified.

Along with sightseeing in Chandigarh, we rented a rowboat, and my friends taught me how to row. It seemed tough in the beginning, but in about an hour, I got the hang of it.

In Chandigarh, we saw a lot of Sikhs. Sikhs make up approximately 10% of all ranks in the Indian army and approximately 20% of its officers, even though Sikhs form only 2% of the Indian population. According to the 9th Sikh Guru, Tegh Bahadur, the ideal Sikh should have both Shakti, and Bhakti. This was developed into the concept of the Saint Soldier by the 10th and final Sikh Guru, Guru Gobind Singh. Protecting the religious and political rights of all people and preventing discrimination is an integral part of the Sikh faith.

The 5th Guru Arjan Singh was martyred by the Mughal ruler Jahangir for refusing to convert to Islam. The martyrdom of Guru Teg Bahadur (The 9th Guru) while protecting Hindus from religious persecution is another example of Sikhs sacrificing greatly to uphold religious freedom. Guru Teg Bahadur gave his life to protect the right of Kashmiri Hindus to practice their religion. They were being forced to convert to Islam by Aurangzeb. There was a tradition in Punjabi Hindus to donate one male child to Sikhism, which was their way of paying their debt to Sikhs for their sacrifices protecting Hindus.

The five items all baptized Sikhs are obliged to wear as commanded by the tenth Sikh Guru, Guru Gobind Singh are Kesh (uncut hair usually kept under a turban), kanga (comb), kachha (brief/underwear), kara/kada (an iron bracelet, which can be a defensive weapon), and kirpan (an iron dagger, knife or sword for defense).

Months leading up to the partition of India and following it in 1947 were marked by heavy conflict in Punjab between Hindus and Sikhs on one side and Muslims on the other. The effect was the religious migration of Punjabi Hindus and Sikhs to India's portion of East Punjab, now just called Punjab.

We visited Chandigarh while India was in war with China over border disputes. There was a great surge of patriotism all over India, but I personally knew very few in Bombay who had a member of their family enrolled in the armed forces. However, in Chandigarh, almost everyone talked proudly about a member of their family or friend's family who was serving in the armed forces. Some of them were posted to fight in the war with China. I was touched when I talked to these families. I felt proud of them, but I also felt a little ashamed that I had not done the same.

Deharadun/Rhishikesh:

We visited the Indian Military Academy in Dehradun. Being among these uniformed soldiers was a unique experience. The alumni of this academy had fought in every war and produced many distinguished soldiers and leaders.

We went to Rishikesh, which is named after one of many names for Lord Vishnu. We performed an arti at one of the temples. Lord Rama came to Rishikesh to give penance for killing Ravana. Lakshmana came with him. Both crossed the river Ganges. Ram Jhula and Lakshman Jhula, two pedestrian bridges, are named after their crossing. We crossed a suspension bridge. This bridge had jute ropes as the main catenary element of the suspension bridge. As I crossed the bridge, the bridge was shaking horizontally and also bouncing vertically. I was nervous and I sincerely hoped that a competent engineer had designed it. If not, I could drop several hundred feet into the cold Ganges water. The bridge did not fail, and I survived. To celebrate and express my appreciation to the gods, I took a dip in the icy cold water. The water was sparkling clear—deep blue and green, unlike any other body of water I had ever seen.

Nainital:

The Nainital Lake is one of 64 religious sites where parts of the charred body of Shakti (Parvati) fell on earth while being carried by Lord Shiva. The spot where Sati's eyes (or Nain) fell came to be called Nain-tal, or lake formed of the eye. The goddess Shakti (Parvati) is worshipped at the Naina Devi Temple.

Nainital is a popular hill station at approximately 9,000 feet elevation. Normally people go to Nainital in the summer to cool off. But we were there at the time of Diwali, right at the beginning of winter. All of us who lived all their life in Bombay, where temperatures rarely fell below 60 degrees F, were not prepared. All we had were light sweaters. We walked around the lake at night for a long time, even though the temperature was near freezing. It did not snow while we were up, but we did see a little bit of snow on the ground next morning. That was the first time I had seen snow in my life. There was not much snow on ground, but enough to make small snowballs and play with it.

As the day progressed, the temperature warmed up. We rented rowboats. I had already learned the basics of rowing in Dehradun. So here I just perfected it more. It was fun.

This lake was a very favorite place for Bollywood moviemakers. It was a hill station, so film crews enjoyed filming there. There was lot of beautiful scenery of the Nainital mountains, Nainital lake, and a view of Himalayas that were shown in many movies. In fact, on the day we were there, an area of the lake had been closed off for shooting. They shot in the bright daylight and used floodlights also. If they needed to film a night scene, they just used dark filters to make day appear as night.

Allahabad/Prayag:

Benares/Varanasi is recognized as the oldest living city in India. Allahabad is next. Allahabad's original name, Prayag, means place of offering. People came to make religious offerings because three holy rivers—Ganga, Yamuna, and Sarasvati—merge here. The merging area is called Triveni Sangam. The name Prayag is referenced in Vedas as the location where Bramha, Creator of Universe, attended a ritual sacrifice. It is also mentioned in Rigveda and Puranas. Lord Rama spent some time here. The physical location of the city has changed over time due to the shifting of the Ganga and Yamuna, and the disappearance of the Saraswati river.

Starting in the late 19th century until the independence, Allahabad was a revolutionary center of the Indian National Congress. The Nehru family homes of Anand Bhavan and Swaraj Bhavan played important roles in the revolution. The first ideas of the Pakistani nation also were born here when the All-India Muslim League proposed a separate Muslim Nation. Nityan Chatterjee and Chandrashekhar Azad became martyrs for bombing the British club here. Allahabad had strong connections with seven of the fifteen Prime Ministers of India. Besides Jawahar Lal Nehru, Indira Gandhi, and Rajiv Gandhi—all from the Nehru family, Lal Bahadur Shastri, Gulzarilal Nanda, V.P. Singh, and Chandra Shekhar were born and received education in Allahabad.

We visited the Nehru family homes of Anand Bhavan and Swaraj Bhavan. Anand Bhavan was built by Motilal Nehru and served as the residence of the Nehru family, and was converted into a museum focusing on the Nehru family. A previous Nehru home became Swaraj Bhavan which was headquarters for the Indian National Congress.

We also took a ferry ride to see the actual mixing of the two rivers, the Ganga and Yamuna, and the third missing/invisible Saraswati river at Triveni Sangam. One had a bluish gray color, while the other one had a dark green color. This is also the site for the famous Kumbh Mela, celebrated every twelve years, where hundreds of thousands of people come to participate in religious activities.

Benares/Varanasi/Kashi:

Varanasi, the oldest living city in India, a city on the banks of the Ganga, is a major religious hub of great importance for Hinduism, Jainism, Buddhism, and Sikkhism. Buddha gave his first sermon in Sarnath nearby. Adi Shankaracharya popularized Shiva worship here. Tulsidas wrote Ram Charit Manas here. Kabir and Guru Nanak visited Varanasi. Recently, Narendra Modi chose to run from Varanasi to become the Prime Minister in 2014. Hindus visit the city and take a dip in the Ganga water to get their wishes fulfilled, wash away their sins, and achieve Moksha. For these reasons, we did the same. We visited some of the ghats. We offered flowers to many gods. We took a dip. We floated paper diyas at night. We took a sightseeing boat ride.

Kolakota/Calcutta:

Calcutta, the city's name until 2001, and the third largest city in India, after Delhi and Mumbai, is located on the East bank of the Hooghly River. Calcutta served as the main entry port for the East India Trading Co. It also served as the capital for British Raj until 1911, when the capital was moved to Delhi. Calcutta was major center of the Indian Independence movement. However, after the independence, due to a number of reasons, Calcutta has stagnated economically.

The partitioning of India led to a large number of Muslims leaving for East Pakistan, and a large number of Hindus arriving from East Pakistan. A violent Naxalite movement since the 1960s has hurt the area. The independence of Bangladesh led to a massive influx of hundreds of thousands of refugees starting in 1971, and starting in 2017 Rohingyas from Myanmar. State governments, including the longest serving democratically elected Communist Party and subsequently the Trinamool Congress, have provided bad governance that has continued the stagnation.

Four of the five Nobel laureates from India have strong connections to Calcutta. They are Rabindranath Tagore, Literature; C.V. Raman, Physics; Amartya Sen, Economic Studies; and St. Mother Teresa, Peace.

We visited the city. We walked and drove over Howrah Bridge. We also took a trip to Shantiniketan to be the place where Rabindranath Tagore ran the university after his father.

Darjeeling:

When we went to Darjeeling, we had to change trains to a narrow gage mountain train at Siliguri. Portions of the ride were on a steep incline. Both when going up and coming down, many of us had to get off the train to reduce the load on the engine going up and reduce braking force when coming down, to keep it under the capacity of the equipment.

Darjeeling is a very popular tourist destination. The city, located at about 7000 feet above sea level, has a cooler climate and beautiful scenery. The name Darjeeling, derived from the Tibetan word for thunderbolt, Dorje and for place, Ling, means place of the thunderbolt baton of Lord Indra.

Darjeeling is famous for tea grown in this hilly region with cool temperatures and lots of rain. We had a classmate who was from Darjeeling. Even though he was not traveling with us, he made arrangements with his dad to help us on our trip. When we visited him, we were surprised to find out that his dad was very rich and owned a huge tea estate. He provided us with a car and driver. He arranged for us to take a tour of his estate, where tea was picked and processed for domestic sale and export. The price of tea depended on how big the leaf was. And different sized leaves were processed under different standardized trade names.

Darjeeling has been politically intertwined with four countries that

TRIP WITH FRIENDS

have all laid claim to it. Nepal/Gurkhas, Sikkim, Bhutan, and finally as part of East Bengal in India. Our friend was of Nepali Descent. The fight for Gorkhaland was finally settled with concessions of more Gorkha representation in local government.

I am Marwari. Marwaris are famous for locating business opportunities. Even so, I was surprised to see so many Marwari shops in downtown Darjeeling.

Most roads in Darjeeling were sloping. Bombay, which is fairly flat and at sea level, had not prepared me to imagine that we would be walking on a sloping road among the clouds. But that is exactly what we did. As we were walking, I spotted a car that seemed to have been parked on the roof of a building. I was puzzled at first. But then I remembered that in our hotel, where we had rented rooms on the ground floor, we discovered when we walked to the back side of the hotel that we were on the third floor. I was surprised to see people refer to places as up or down. Also, just because you can see something close by, it does not mean that it is close by. You probably have to walk on a road that winds back and forth. So walking/driving distance can be shockingly large compared to the distance the way a crow flies.

Darjeeling is a beautiful place with lush, hilly terrain. You can easily see Kangchenjunga, part of Himalaya and the third tallest mountain in the world. On a clear day, you can see Mount Everest, the tallest mountain at about 29,000 feet.

Our friend's driver drove us to Tiger Point. We started at about 2 a.m. We were freezing, but the driver seemed very comfortable. Obviously, he was used to the weather. But he also knew to wear layers of clothing that he would peel off as the day progressed. We had never thought of layering in Bombay, so we had nothing to peel off. We froze until long after daybreak. The driver drove early in the dark morning through the mountainous region. We reached Tiger Point at about 4 a.m., about a half an hour before the first rays of sun hit the top of trees. When this happened, it was magical.

First, just a sliver of gold line in the far distant trees. Then the line became wider and the line approached us. Slowly, the sun covered the tops of all the trees. By 6 a.m. from our high vantage point, everything was covered with sunlight. The show was over. We ate breakfast and headed back to town. Now it was daytime. We could see how treacherous the road was that we had just traveled on earlier in the dark: A big drop on one side, one lane, sharp curves, poor sight distance, and the fog from the morning. But we made the trip safely.

We went to Kalimpong. This was at a time of the Indo-China War of 1962. We had heard that there were lot of Chinese spies in Kalimpong and Gangtok. We walked to the market suspiciously looking at everyone with a Chinese look to determine if they were a spy. We could not spot any spies!

After this, we went to Gangtok, which was capital of then independent monarchy of Sikkim. No visas or passports were required for Indian citizens. Our Indian look was enough. It was made a state of India in 1975. We visited a Buddhist monastery and a king's palace. I was surprised to note that women were working in labor jobs with their babies strapped to their backs, while men were drinking tea, smoking, chatting and relaxing. This was our last stop. It was time to return. We went back to Darjeeling, then to Calcutta, and then to Bombay on a 1200-mile, 48-hour return train trip.

During the 17-day trip, we had seen and experienced so much. We knew we would never forget that trip. As I am writing about it 57 years later. I want to thank you, Kanti, for arranging every detail so well, and foreseeing that we would never make such a trip again.

CHAPTER **27**

India Defense Fund and Prithvi Raj Kapoor

DURING OUR TRIP, we had visited Dehradun and Chandigarh. In Deharadun, we had met army personnel. In Chandigarh, I talked to some individuals. Many of them had family members in the army, in the Border Security Force and/or in combat in the war with China, that was going on while we were traveling and having fun.

When we came back, we resumed our normal college life. A normal part of our college life was to see Bollywood movies. My close friend, Dipak Shah, and I were waiting in line to get our tickets for a movie. Suddenly, I had a strong guilty feeling inside me. I described it to Dipak. I said, "You know, while we are waiting in line to see a movie and have fun, Indian soldiers are dying on the battlefield in war with China. It does not feel right." He agreed with me. He said, "But what can we do?" I had no answer. We got our tickets, watched the movie, and went home. But the guilty feeling did not go away. I knew, I had to do something. I thought that raising money for the India Defense Fund would be a positive action. I went to one of my friends. I told him that we needed to raise some money for our soldiers and India's defense. His response was, "No one is going to give you any money. Besides, what difference will it make?" I was discouraged.

The guilty feeling would not go away. So, I decided that I needed to explore the idea with another friend, Ashok. He was very enthusiastic. He said, "Why not?" With his encouragement, I came up with a plan working with Dipak, who now was "all in". We picked a day and called it "India Defense Fund Day". On this day, everyone who ate at the college canteen would be paying four times the menu price. The extra money charged would be donated to the India Defense Fund. When I informed my classmates, there was wide support. I checked with the canteen management, and they were ok with it.

Suggestions, some serious, some not, started pouring in. One such suggestion which was made in jest appealed to me. The suggestion was that the very few girls that we had in our college would serve as waitresses for a short time on that day. I liked the suggestion and implemented it without consulting the girls. I assumed that their consent would be received for such a noble cause without any difficulty. Posters were displayed at a number of places to publicize the fundraising. Girls with smiling faces were drawn on these posters. In all the excitement of the upcoming event and preparations for it, I forgot to talk to the girls about it.

On India Defense Fund day, everyone willingly gave the extra charges to eat in our canteen. As the day progressed, we thought it was time for girls to show up for a little bit. We approached the girls to inform them that it was time for them to help us. Unfortunately, the first girl who I talked to got very upset. She refused to help, explaining that we had not treated them with respect. She said, and I realized immediately that she was right, that we should have taken their consent, and not assume it, before putting up posters. We apologized, but she did not budge. The other girls supported her position. Dejected, we walked back to the canteen. I was really worried and ashamed. I was even more worried how we were going to explain, if someone asked, about the girls.

INDIA DEFENSE FUND AND PRITHVI RAJ KAPOOR

Suddenly I had an idea. I turned to Dipak and said, "Why not Prithvi Raj Kapoor?" He said, "What do you mean?" I said, "If Prithvi Raj Kapoor came, we will raise more money, and everyone will forget about our fiasco with the girls." Dipak was such a friend that he would support me no matter what. Without hesitation, he said let's do it. I said, "Right Now? It is 2 p.m." He said, "Yes, Right NOW. We don't have much time." We walked over to Prithvi Raj Kapoor's apartment, which I passed by every day on my way to college. It was only two blocks from college.

Bollywood Icon Prithvi Raj Kapoor on a 2013 Postage Stamp of India

Prithvi Raj Kapoor was born in 1906 and died in 1972. He was a pioneer of Indian theatre and of the Hindi film industry. He started his career as an actor, in the silent era of Hindi cinema. He starred in *Alam Ara*, a silent movie with Ardeshir Irani, Master Vitha, and Zubeida in 1931. He acted in movies until his death.

GROWING UP IN MUMBAI, INDIA IN 1940S, '50S AND '60S

He is best remembered for his role in *Mughal-e-Azam*. *Mughal-e-Azam* (a Mughal emperor) is a 1960 Hindi film directed by K. Asif and produced by Shapoorji Pallonji. It stars Prithvi Raj Kapoor as Emperor Akbar, Dilip Kumar as his son Jehangir, Madhubala as a fictitious character named Anarkali, and Durga Khote as Hindu wife of Akbar. The movie is about the love between Mughal Prince Salim (who became Emperor Jahangir after Akbar's death) and Anarkali, a court dancer. Emperor Akbar disapproves of the relationship. Eventually the conflict leads to a war between father and son, and the death of Anarkali by live burial in a stonewall. The production of Mughal-e-Azam began in 1944, and the movie was released in 1960 after 16 long years of lavish production. K. Asif, who directed this movie, may be compared to Cecil B Demille who directed *The Ten Commandments*, *Cleopatra*, *Samson and Delila*, and many other epic movies, all of which I saw while growing up, just like Demille, K. Asif's movies had a huge budget and achieved great success. Mughal-e-Azam had the widest release of any Indian film up to that time, and patrons often queued all day for tickets. It broke all box office records in India up to that point and became the highest-grossing Bollywood film of all time, a distinction it held for 15 years.

Prithvi Raj Kapoor was chosen for this role because of his theatrical ability, his strong gravelly voice, his height, his impressive build, and strong personality. By the time of this movie, he had already become a patriarch of the most successful dynasty of film actors and producers. His three sons—Raj Kapoor, Shammi Kapoor, and Shashi Kapoor had all become famous as heroes in many movies. Raj Kapoor had also produced many movies under the trade name RK Productions and most of them were hugely successful. Many of his grandchildren and great-grandchildren also grew up to become hugely popular actors and actresses.

Can you believe that two of us were going to walk over to this icon of Indian movies at the peak of his career and try to meet him in his

INDIA DEFENSE FUND AND PRITHVI RAJ KAPOOR

apartment without any invitation or appointment, invite him to come to our college with no notice so that we could help raise a small amount for the Defense Fund in our college canteen, to save our face in the process?

It was stupid. But sometimes it pays to be stupid. How else would you try something so ridiculous? We walked up to his apartment building. We asked a neighbor where he lived. He lived on third floor with his youngest famous actor son Shashi Kapoor. Luckily for us, there was no security guard on duty to stop us. We walked up to the third floor. We knocked on the door. So far, so good. The door opened. A servant greeted us. He asked us the purpose of our visit. I explained that we were students at a nearby engineering college, VJTI, and wished to see Mr. Prithvi Rajji. He asked us to take a seat on the sofa in the living room. He asked us if we would like a drink of water. We declined. He asked us to wait and went inside. We looked around. The living room was not as big as I thought it should be. It was decorated fairly simply. In a few minutes Prithvi Raj Kapoor walked in.

He was as impressive as he was in the movies. He greeted us in his strong gravelly voice. We explained the purpose of our visit. He asked us to wait. He came back with an album.

He showed us his picture with Zhou Enlai (previously spelled Chou En-lai), China's Prime Minister and second most powerful person after Mao Zedong (previously spelled Mao Tse Tung), father of last revolution in China. Prithvi Raj Kapoor had led an entertainment troupe to China to promote India China friendship and returned just a few months earlier.

The Indian Prime Minister Jawaharlal Nehru and his defense minister Krishna Menon were trying to maintain good relations with China. India was socialist and democratic, and Menon was a communist and democratic. Both were promoting friendship between India and

China. I was in the crowd of 25,000 people at Brabourne Cricket Stadium when Mr. Nehru and Chou En-Lai claimed their friendship with a loud slogan "Hindi Chini Bhai Bhai!"/Indians and Chinese are brothers, brothers (brother was repeated twice for emphasis and rhyme).

Prithvi Raj Kapoor expressed to us his shock and disappointment with the Chinese, that shortly after his visit to promote friendship in which the Chinese had expressed friendship, they had attacked India. He said this attack was cowardly because it was a surprise attack, a dishonorable way, similar to stabbing in the back.

War was not going well for India. The Indian army was ill prepared for a war in a cold climate, and in mountains in the Himalayan range, which until then had provided a peace promoting barrier between India and China. The Indians were brave and were willing to undergo great sacrifice to fight this war. Prithvi Raj Kapoor, who was a Sikh, was very patriotic. When he was told that we were raising funds for India Defense, he did not hesitate even one second in encouraging us in our small effort. He told us he would come. We thanked him and returned to the VJTI canteen.

As soon as we got back from visiting Prithvi Raj Kapoor, we spread the word that Prithvi Raj Kapoor had accepted our invitation. No one could believe it. It was already close to 3 p.m. and the canteen was going to close at 5 p.m. So, we would not have him for long, but still everyone was excitedly looking forward to seeing him. We had not set an exact time for his arrival, but we expected him to show up soon. The canteen clock kept ticking and now it was 5 p.m. and Prithvi Raj Kapoor had not shown up. We waited for another 15 minutes, and then we gave up. Disappointed, we settled the account with canteen, collected our share and headed to the exit. Before we got to the exit, someone screamed "Prithvi Raj Kapoor is here". We turned, and indeed he was there. He had entered through the kitchen back door. Maybe that is how they do it in Bollywood. Everyone returned. There were no customers, but there were about 15 volunteers and few staff members still there.

We welcomed him. We pulled a few tables together. Everyone selected something to eat. We sat there talking to one of the greatest of Hollywood movie actors for about a half an hour. He told us many stories; he answered all our questions. He even asked questions of us. Then he left through the kitchen back door, the same way that he came in.

We stayed back. We sat there for a while, stunned. We could not believe what had just happened. None of us would ever forget the meeting with Prithvi Raj Kapoor. We had raised about 1,000 Rupees ($200), which was donated to the India Defense fund. We had a good day.

CHAPTER **28**

Professional Career Begins

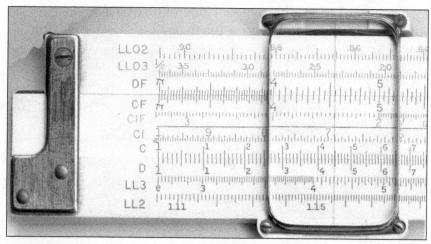

Slide Rule was a big time saver over use of logarithmic tables that Author used in Science College, which replaced manual methods of high school.

AFTER OUR TRIP to about a dozen scenic, historic, and famous places, we settled down to a normal college life. There was much greater emphasis on studies now. For the next one and half years, I was totally devoted to studies. Your performance in the final exam of final year was important. First you were trying to prove to yourself that you are capable of handling difficult problems. Your performance in these examinations was also likely to affect your first job and maybe your

PROFESSIONAL CAREER BEGINS

entire future. So everyone worked the hardest in this final period. My older sister tried to help me by getting up early and making hot tea so I could stay awake and gain a few more study hours early in the morning. But after drinking hot tea that she prepared so lovingly and at such inconvenience to her, I kept going back to sleep right after finishing her hot tea. After it happened several times, she gave up. Fortunately, even without this extra study, I did fine.

We were also starting to plan our professional life after graduation. Some of my friends, including Kanti, were very clear. They knew that they were going to go abroad for graduate studies. Their families were very supportive, and they were in a strong financial position to support their study and travel abroad. Therefore, they started applying for admission to American colleges during the final year of college.

I was not sure about the financial support from my parents. One reason was that Baiji was strongly opposed to my traveling abroad. My parents had heard of some cases where a boy had gone abroad, married abroad, and never returned. Baiji was not willing to take that chance. All parents in India raise their sons to take over the financial responsibility for the family as soon as possible and as much as possible. Sons who travel abroad may escape that responsibility and leave their parents without such support. I did not want to escape this responsibility. I therefore decided that I would not go abroad. I had seen many engineers who had succeeded in their professional career in India. There were some who had succeeded in their own consulting business. I felt that I should be able to do the same. I therefore decided not to apply for admission to any foreign university.

I finished second year of engineering at the top of my class. A year later, after my exams for the final year of engineering were finished, everybody took their last long vacation before exam results were announced and were officially awarded a college degree. While I was waiting for the results, I applied for a job as a structural engineer in my own unique cheap way by sending handwritten cover letters including a brief resume to some consulting engineer doing structural engineering work. This information was squeezed onto a postcard that I mailed. My method worked. I got several interviews.

One of the companies that interviewed me turned me down for a strange reason. They readily agreed that I was well qualified for the job, but the chief engineer who interviewed me was sure that I would not stick around long. He was sure that I would definitely go abroad for study considering my strong academic background. My reassurance, to the contrary, did not convince him otherwise. As it turned out, he predicted my future better than I could. I eventually did leave India.

I got my first job offer from a Kane consulting engineer company, which I accepted immediately. The company consisted of the owner, who was a structural engineer, and a staff of about six engineers, three draftsmen and a secretary, all of whom were males, except the secretary. I worked hard at this company and designed concrete framed apartment and commercial buildings. I got to a point where I felt that I could start my own consulting company doing the same.

I talked to a friend, Sampark, who was also my classmate from VJTI, to go in partnership with me. He readily agreed. Mr. Sampark was gracious enough to allow my name to be the first one in our business name "Kothari and Sampark, Structural Engineers". We printed visiting cards, in black and white, in minimum quantity. Both of us kept our jobs, while we visited a number of architects, and tried to get work from them to get started.

While we were waiting for our first project, with inspiration from my mentor friend, Surendra, I obtained a survey project for a fee of Rs. 50 ($10). I had to work really hard to earn this Rs. 50. That, and the lack of progress in getting consulting work after a few months was beginning to weaken my resolve not to go abroad for studies.

PROFESSIONAL CAREER BEGINS

While I worked at Kane, I had applied to a very large consulting engineer company, Dawawala Engineering Co., and had not heard from them. About three months after I started work at Kane, I decided to stop by and inquire about the status of my employment application with Dawawala. I went to the company at about 8 a.m., so that after I got done, I would go to my job at Kane in time. When I arrived at Dawawala, before I could ask any questions, the secretary directed me to join a group of well-dressed young men sitting in a room. I complied. In a few minutes, a gray-haired gentleman walked in, introduced himself as the chief engineer, and asked us to proceed to another office. Again, I complied. They handed me some forms to fill out. I suddenly realized that I had been hired by Dawawala.

I was expected to fill out the forms agreeing to terms and conditions of employment and start my employment that morning as an engineer with a salary of Rs. 250 ($50) per month, with a promised raise to Rs. 350($70) per month in six months. I was still employed by Kane. My lunch tiffin was scheduled to arrive at Kane. And here I was expected to begin a new job with Dawawala. I was shocked at the job offer, but I went along.

I worked the whole day at Dawawala. After work, I went to Kane, stayed outside their office exit door and waited. The secretary came out. I stopped her and I told her what had happened. I explained to her that I am ashamed at my conduct. I told her I hoped that everyone would try to understand. In any case, I asked her to inform and apologize to Mr. Kane on my behalf. I handed her whatever belonged to Kane. I did not ask or receive my last paycheck. Thus ended my first job.

The work environment at Dawawala was quite different. It was a big office with several hundred engineers working on very large industrial projects in modern airconditioned offices. Everyone wore a tie. We had very large tables, and a lot of space between us. I worked on an industrial steel producing project. The work that I did at Kane was much less challenging and much more repetitious. I felt that I learned quite a bit already, and that there was not as much left to learn. The work at the Dawawala designing industrial facility was much more challenging and varied. I was happy to have changed the job, even though I have regretted how it happened.

My project engineer was Mr. Patel, who had completed his master's degree in structural engineering from the University of Illinois at Urbana/Champaign. He looked smart. He had a very good personality. He spoke good English with an American accent. He was very knowledgeable and introduced American codes and technologies in our work. I was impressed by him. He earned RS. 800 ($160) per month. I said to myself it I can triple my income and improve my personality and stature by going to the United States for three years, it is not a bad deal. Mr. Patel encouraged me to go abroad. He encouraged me to talk to some more persons who had been abroad, learn as much as I could, and then make my decision.

CHAPTER 29

Student Visa for United States

THIS WAS THE pivotal moment for me. I started considering changing my mind about not going to the U.S. for further studies. I began applying to schools in the U.S. for graduate study. I applied to about twenty universities for graduate study. Baiji was against it, but I did try to calm her down. I told her that I had not made a final decision about going abroad yet. I had not. I was just trying to keep my options open. She had no choice except hope that I would not go.

While I was waiting for things to sort out, I started preparing myself for a high possibility that I may be leaving India soon. In order to help with educational expenses, I thought that I should acquire some additional skills. In India, very few women took up employment at that time and most secretarial jobs were handled by men. I thought that if I could find a job as a temporary secretary, I would be able to supplement my funds for education. So, I took a short typing course. As it turned out, I never took a secretarial job in United States. However I did benefit greatly when I had to do my own typing and when we all started using personal computers. I am one of the few Indians who use both hands.

I had enjoyed music and could sing a little better than average person. I knew that this may be my last chance to learn Indian music. So,

I joined an evening music school. I tried my hand at harmonium, but my patience ran out and I quit in a few months.

Applying for admission was a long, tedious, and expensive process. First, the communication with American universities was very slow. After you wrote your letter requesting application forms, it could be weeks before you would receive a response. My letter was sent by airmail, which took two to three weeks to reach a school. The school would send a heavy part of the admission package back by sea mail, which would take six to eight weeks. It took a while to fill out the forms. And then, since Xerox machines were not available at that time, the original signed application was sent, but no copies were kept. All my marked sheets and certificates had to be typed and certified as a true copy signed or stamped by a school authority. As the applying process was progressing, I was getting more convinced that going to the U.S. to study was a good idea.

I now began preparation for my trip in earnest. First, I found a travel agent who would help me to obtain a passport. This simple step took so many trips to various offices, waiting for hours to see the right person, and to be told again and again to comeback in a few weeks, or to see someone else. It took months, not weeks, to clear police investigation. After months of frustrating efforts, I finally obtained my passport.

After a few months of applications, several schools accepted me and sent me their I-20 form, which certifies that I have received admission to their school. You had to have this form to apply for an F-1 student visa, and to obtain Reserve Bank of India clearance to travel and to transfer funds for education. I talked to Mr. Patel and few others. Considering their input, I chose the University of Michigan in Ann Arbor, Michigan. Tuition was reasonable, and the school had a good reputation.

STUDENT VISA FOR UNITED STATES

I took the English Proficiency Test devised by the University of Michigan. I was approved for a visa. And I did not need remedial English. I took a blood test in which nothing objectionable was found. After I proved with documents that I had strong ties with India, and after I swore that I had no intention to settle in United States, I was granted a student visa. I wondered whether a white student coming from Europe would be asked the same question about his commitment to return to his country after his studies were completed. My feeling was that he would not have been asked that question. While getting a visa for most people is very torturous, in my case the process was not bad. Welcome or not, when I got my visa, I considered myself very lucky. The day I got my visa was one of my happiest days.

I applied to the Reserve Bank of India to get foreign exchange for my studies. Again, after going through a lot of paperwork, including getting signatures from a Justice of Peace, generally a judge or a respected politician, and proving my financial resources to pay for my education and travel, I got foreign exchange clearance. Officially, I was given US$8 cash for travel expenses until I reached the U.S. and could arrange for the transfer of money through banking channels. However, I was also able to arrange for some traveler's checks, unofficially, which I would cash in the U.S. for expenses until money was transferred for my education and living expenses. I had all the paperwork ready.

Now the decision to pay for travel ticket was upon us. I had saved some money, but it was not going to be enough to pay for airline ticket. Travel by ship, which would take approximately 21 days, would save money. My parents had accepted my decision to go to USA, and considering that I was going to be married on May 26, 1964, less than three months before my departure, found a way to come up with the difference, so I could spend more time with my wife before departing.

CHAPTER 30

Arranged Engagement

BUT NOT SO fast. After I joined Dawawala, Baiji noticed a change in my attitude about going to the U.S. She saw how hard I was working to apply for admission to American universities. She was not happy. She wanted me to stay and take care of the family. She wanted me to join Babuji's business as soon as I finished college. I was not interested. I told her that even if I go abroad, I was planning to go for about three years, and then come back in a better position to help the family. She was not convinced. She had one more concern. She was afraid that I might get attracted to an American girl, get married to her, stay permanently in the U.S., and abandon them. I told her that I had no such desire. But she was not convinced. Even though I had not even secured an admission to any American college, she started spreading the word that she was in a real hurry to get me married before I left India.

We as family often used to visit a close family friend, Dhariwals, who lived in the Kemp's Corner area of Bombay. We were close family friends. Our families used to visit each other quite frequently. I used to visit them often on my own. Mrs. Dhariwal mentioned to Baiji about a girl from Dhulia, Maharashtra, which is about 300 miles from Bombay. Baiji showed me her picture and asked for my reaction. I looked at her picture and returned it with a comment "she looks

ARRANGED ENGAGEMENT

good". With this preliminary approval from me, families exchanged horoscopes and talked to each other on phone.

Great majority of marriages in India are arranged only after an expert astrologer has compared the horoscopes. The idea is to avoid marriages that may end up with serious marital conflicts, health and/or wealth problems, or the early death of a spouse. If a bad report is received, sometimes a second opinion is sought. The marriage will probably not take place if the problems persists. A clean report does not always end up in a blissful marriage. You just have to wait to find out. By the time you find out, the astrologer is gone.

Our horoscopes did not indicate any problem. Mrs. Dhariwal and Dhulia girl's bhabhi/brother's wife were both from Indore, Madhya Pradesh. Mrs. Dhariwal knew and learned a lot about the girl and made a strong recommendation to Baiji for Dhlia girl's marriage to me. Baiji was interested.

I am not sure if it was arranged or not, but the Dhulia girl was visiting the Dhariwal family. The Dhariwal family suggested that I visit them and meet the girl. I was already intensely working on getting admission to an American college. It was 1963 summer; I was only 20 and I had no interest in getting married just yet. So, I told Baiji why waste time? Baiii turned around and counter-argued: why not? Out of respect for her, I said, "Ok, I will visit the Dhariwals and meet the girl." I cautioned her that I had no interest. Baiji said that was fine. Baiji informed Mrs. Dhariwal that I would be coming the next morning to visit.

I took the familiar train and bus route to their home near Kemp's corner and the American embassy, in the area where India's richest man would later build a billion-dollar mansion, which would be the biggest home in the world. I sat down in the living room like I had a done many times before. They offered me some snack and water. Mrs. Dhariwal walked in with girl from Dhulia (soon we will all call her Dhuliawali.

GROWING UP IN MUMBAI, INDIA IN 1940S, '50S AND '60S

Or Dhulia Girl or "The Girl" for short). She introduced me to the girl. She said, "You guys talk. I need to do something in the kitchen."

I was struck by this girl's beauty, her slim body, her smile, her long hair, her fair skin, and everything about her. It was the moment of love at first sight. There were just two of us in the living room. I was not sure what to say. She probably was taught not to speak first. So, no one said anything for a few seconds. I realized that I was going to have to speak first. So, I asked her some questions about her studies. She replied. The ice had been broken. It was now comfortable talking to each other about our likes and dislikes, our interests and hobbies, and our families.

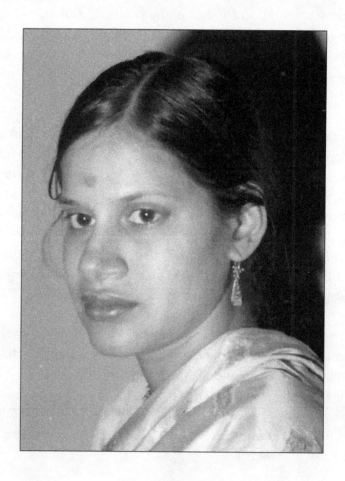

Author And Wife Sarala Then and Now

I felt I could talk to this person for a long time and keep enjoying the conversation for a long time. Based on her response, it seemed to me that feeling was mutual. After about half an hour, Mrs. Dhariwal came and gently broke up our meeting. Sarala went back in the kitchen. I stayed back for a little while talking to Mrs. Dhariwal. I always talked with her a lot whenever I visited them. Then I left to go home.

My trip back home was like no other trip before this. I was so happy to have met this beautiful girl. She was more beautiful than anyone I had ever seen before. When I went to see her, I had no intention of getting married. Now, I was repeatedly thinking, *why not?* I was thinking of how I would manage marriage and my visit to the U.S. I was happy. I was on cloud nine, whatever that means. Until then, I did not know how it would feel to be on cloud nine. Now I knew. I kept remembering and going over every word we spoke: every gesture that she made, every smile, a slight laugh at my comments, and her interest in my silly questions. I was trying to impress her with my smartness, my sense of humor, and my interest in small joys of life. It seemed that she was impressed. I could not be happier.

Even though there were some small regrets for not having made perfect choice of words or expressions, I was generally happy of the way our meeting went. I was sure that this Dhulia Girl would like me and find me to be a good match; and if I informed Baiji that I was ready to get married to this girl, our marriage would be arranged. As I returned home, I was not aware of anything around me. All I could see was her—not the sidewalk, not all the people that were going by me, nor those people and poles that I kept running into. I don't know how I made it back home without getting hit by a car, a bus, or a train. I walked to the station, took the train, and came home. But all through this, my mind was busy thinking about only this girl from Dhulia—nothing else. Baiji asked me how I liked the girl. I said, "She is fine." They could tell that I loved the girl. I did not have to elaborate.

ARRANGED ENGAGEMENT

Later that evening, I met my friends for my usual walk to Santacruz Juhu Beach. That day, I did most of the talking. My friend would ask a simple one- or two-line question. And I would respond with a long answer. It took me almost four hours to describe all the details of a thirty-minute meeting with this Dhulia Girl. I am sure that I repeated the same story more than once just so I could add an additional detail that I may have missed or a slightly different version of details as I recalled. I was so enjoying describing my meeting with the Dhulia girl that I wanted to be sure that I did not miss anything. Everything that happened in those thirty minutes was so important, so different, and so pleasurable. I could not fully put in words what I felt, but I kept trying, all evening. Among our friends, I was the first one to have an experience of meeting a girl for marriage. Someday, they expected to get their chance. So, they were very interested. And I was so happy to share all the details.

Even though I was very impressed with this Dhulia Girl, marriage was not my high priority at that time. I was more focused on my trip to the U.S. After my comments to Baiji that the "Dhulia Girl was fine", even though in my heart I was saying that the "Girl from Dhulia is wonderful", I did not follow up on the subject of marriage with Baiji.

One day we were invited to the Dhariwal family. I did not know, but it was a specially arranged trip. The father of the Dhulia girl wanted to evaluate me further. He asked me questions about my education, about my plans to go to the U.S. for studies, about my hobbies, about my views on politics and world affairs, etc. I was totally surprised when he asked me to hand over my glasses to him. He examined them to see how thick they were. He wanted to make sure that I did not have a very high prescription. He also asked me to stand next to a cabinet. He and others mentally marked my height, which was later measured to determine my height. He wanted to make sure that I was taller than the girl. The girl's father was also making sure with some questions that I was not a person who would lie or misrepresent.

I found out later that the Dhulia girl's parents had been trying to arrange a marriage for her for about two years, since she was fifteen, which was not very common at that time. During this period, she had been shown to number of boys. But this girl was picky, and she rejected everyone for one reason or another. No one ever told me about this. If they had, I might have been reluctant to see her at the Dhariwal home. The girl's father was determined to make sure he checked everything thoroughly so that there would be no justifiable reason for her rejection. This time she did not reject. Her father was able to convince her that I was a good match, even though this convincing did not happen right away.

Baiji informed me a couple of months after the girl's father's visit that a date had been set for my parent's travel to Dhulia to carry out the "Shagun/engagement" ceremony. This ceremony would mark consent and commitment by both families for my marriage. I got engaged without proposing, and without being present for my own engagement!

Shortly thereafter, my fiancée, Sarala (the Dhulia girl now had a name), came to Bombay for a visit. On that day, I was allowed to meet her. We spent the whole day together: going to restaurants, shopping centers, and Chowpatty Beach. We were chatting constantly. I was doing most of the talking, and she seemed to be interested in every word. We were both interested in knowing each other. I may have fallen in love with a strange girl when I met her for the first time at the Dhariwal home. Now I fell in love with Sarala, my future wife. I was amazed at her beauty, her voice, her smile, her laughter, and everything about her. I was saying to myself again and again, "How can I be deserving her?" That day, we also stopped by a photo studio and got our first picture taken together. I was 20. She was 18.

CHAPTER 31

Arranged Marriage

AFTER THE SHAGUN/ENGAGEMENT had been done, my parents gave me an opportunity to spend a few hours with Sarala, my fiancée, away from family. I enjoyed every moment of it. Now, my parents were assured that everything was going good between us. They asked me if they should set a date for the marriage. I did not hesitate and said, "Sure."

There are many things negotiated between two parties in an Indian arranged marriage prior to engagement and prior to marriage. It is a long list for the girl's side, which includes dowry and gifts from the girl's family to the boy and the boy's family, gifts for the groom and the groom's guests, the number of Baraat guests who will attend marriage, "standard" of marriage, and many other details. There is not much expectation from the boy's side. Non-monetary aspects are also agreed upon, including the details of traditional customs to be honored, and the details of religious ceremonies to be performed. These details are important so that each side can estimate marriage expense, which is the single biggest expense for many girl's family in their lifetime.

I was influenced by modern ideas and laws that discouraged and legally prohibited dowry and other indirect ways of extracting dowry.

However, at my age, and with my upbringing where parents dominated children, I was not very strong in insisting that my parents follow my ideas completely. I, personally, did not make a single demand. However, I stayed out of any negotiations that my parents conducted with Sarala's parents. I believe that they were influenced by my views, and therefore I believe they were moderate in their demands. By and large, it is my belief that most of the gifts were given voluntarily and in keeping with norms in Dhulia. Sarala's parents had visited us and were familiar with our household. Among the gifts they included was a refrigerator, a king-size bed, and Rs. 10,000 ($2,000) certificate of deposit in their daughter's name.

Based on the horoscope and other considerations, the date of May 26, 1964 was set for the marriage. With only a few months left for the marriage and my subsequent immediate trip to the USA, preparations for the marriage and the trip to USA shifted in high gear. A few relatives and friends were contacted to give them the good news of my upcoming marriage. Word spread very quickly from there.

Everyone from our extended family and friends started contacting us to first find out details about the bride and her family, and then to share their joy of my engagement. There were lots of phone calls; lots of visiting and visits; lots of suggestions; and lots of offers to help.

There was one exception. My family doctor, Dr. Gandhi, who treated me like a son, was disappointed. He thought that I was a fool to get married before going to the USA He told me more than once that I would be worth so much more after completing my studies. I had no interest in dowry, so I just ignored his comments.

My marriage was an arranged marriage. I did not know the girl and had never met her until my first meeting at the Dhariwal residence. But our arranged marriage was modified from traditional arranged marriage to suit modern times and me. Traditionally, a marriage was arranged early in childhood, sometimes even before a child was born. Then a marriage would take place among the two children, even before they were teenagers. Girl would go back home and would be raised there. She would be sent to boy's home after she grew up a little, maybe to age 10 to 12, to begin a married life with her child husband aged 12 to 14. The age crept up a little bit under British rule to a point where my elder cousin sisters did not get married until they were aged 12 to 14, and cousin boys until they were 14 to 16. Things started changing after the independence to a point that my fiancée's parents did not start looking for a match until my fiancée was 15. If a good match had been found, it is possible that she would have been married at 16. My parents delayed arranging my marriage to keep it from disrupting my studies. However, once I graduated, even though I was only 20, they were anxious to get me married. If I had not continued my studies to college, it is highly likely that I would have been married by 18.

Because I grew up in Bombay, my parents recognized that changes were taking place in society, and therefore, arranged for me to get to know the girl better before marriage. So, when Sarala's cousin brother was getting married in a village near Dhulia, I was invited and attended the wedding. While I was standing in the baraat, I noticed that Sarala was really tall. I could easily spot her among all the girls and women with ghunghat/covered heads that were standing in a separate group apart from men. At 5' 5", she was at least 3 to 9 inches taller than most everybody in Dhulia. Now I understood why Sarala's father made sure that my height was measured and approved before my shagun/engagement.

After I came back from marriage and reported on how Sarala's family welcomed me and took care of me, everyone knew that we were getting along well. Now there was no reason to hesitate in proceeding with the preparation for marriage.

GROWING UP IN MUMBAI, INDIA IN 1940S, '50S AND '60S

My parents drew up a list of all the guests who would be attending marriage on our side. The total number had already been agreed to. My side, the bridegroom's party, called "Baraat", was going to travel to the bride's home. All close relatives and friends were invited. Some would come with their children. Some without. Some of my friends also joined in Baraat. We had about hundred persons in my Baraat. Train reservations were made for as many as feasible. Others would travel without reservations.

I needed to get a suit for marriage. Now, I don't know why but it was a tradition for the groom to wear a white sharkskin suit and a colorful Rajasthani style turban. I guess the reason is so he stands out. Because no one else will want to wear a white suit. I made countless trips over about a month to the tailor to get the fit just right. Additional normal clothes were tailored or purchased. New shoes were purchased.

A few days before wedding, guests stared arriving at our home. No one stayed at a hotel. They all stayed with us in our two bedroom apartment. A few of them stayed with our close friends or relatives. There was constant chatter in the house with so many ladies meeting after such a long time. They reminisced about their marriages and their traditions. They talked about their families, friends, neighbors, and anything interesting. Men talked a little less, but with so many together, meeting each other after a year or so since the last wedding, they also had a lot to talk about.

A few days prior to wedding, there was a "Pithi" ceremony. A paste of turmeric powder is applied to the exposed part of groom's body, supposedly to make him "glow", by ladies who were closely related family or close friends. I thought it was silly, but I went along.

In the meantime, at Sarala's home things were hectic. From the day I met Sarala at the Dhariwal residence, discussions with her family, particularly between her parents got intense. After Sarala's father had done enough background checks with everyone who knew me and my family and completed thorough inspections and tests that he personally conducted when he saw me at Dhariwal's residence, he was convinced that he had found the best match for his daughter. Her mom, who did not get all this information firsthand, might have been not so sure, but her opinion was easily influenced by Sarala's dad, so that both of them approved of me.

Sarala was in her junior year in college at that time. She stayed on her own in a girls' hostel in Pune where she attended college studying to get a Bachelor of Arts degree in English. She had spent over two years in Pune and was quite comfortable with her independent lifestyle. She knew for sure that marriage was going to crimp that lifestyle. She was young and did not think that there was any need to rush. She thought her parents should wait, at least a couple years, until she graduated. She had been able to postpone marriage for about two years by pointing out flaws in the other prospective grooms. Many times, she had requested a prospective groom to reject her, and they did so partly to make her happy, but mostly to save themselves from embarrassment.

This time, she was having difficult time. She had not asked me to reject her when she met me or any time after that. If she had asked, I would have done so also. Still, she delayed agreeing to the marriage that her parents pushed for a couple of months.

A decision was finally made by her to accept her father's recommendation. This decision was conveyed to Babuji and subsequently the shagun/engagement took place. A few months after that we met a few times and a wedding date was set. Preparation for marriage began immediately thereafter.

GROWING UP IN MUMBAI, INDIA IN 1940S, '50S AND '60S

Author's Mother-In-Law, Maa and Father-In-Law, Bhausaheb in 1963.

A list of all the guests on both sides was drawn up. Invitations were made by handing out invitation cards and inviting guests in person as much as possible. Otherwise, they were also mailed out. Phone calls were not common and there were not too many phone calls.

A priest had already been consulted and arranged. The wedding venue was to be at Sarala's home. Arrangements were made for Mandap, caterers, and nearby facilities for lunch and dinner and for guests' stay.

Most time consuming was the shopping for clothing and 24-karat gold jewelry for the bride and everyone else. Local shops were visited every day and tailors were hired. But it was also necessary to make many overnight trips to Bombay for shopping.

CHAPTER 32

My Indian Wedding

THE WEDDING WAS set for May 26, 1964. The previous night, we all piled into a train and took the trip from Bombay to Dhulia. We ate dinner on train which had been packed with tiffins. Late into the night, we were singing Bollywood songs or telling jokes or stories. At some point, parents came and asked us to stop and go to sleep.

The train arrived at Dhulia train station. We all took rides in Tangas, a single horse driven carriage, with our luggage to a place with large halls. Men and boys stayed in one large hall. We sat and slept on thin mattresses or heavy blankets/sheets spread on the floor. Women and girls did the same in another large hall. In the morning, everyone was served delicious breakfast and tea. Sugary drinks and tea were served periodically. An elaborate lunch was served in a large tent with everyone sitting on the floor on a mat. Food was served in stainless steel thalis/plates with raised edges and katories/small cups. Water was served in stainless steel glasses. Food was carried in large plates, and fluid items were carried in large buckets and poured with stainless steel bowls/cups.

Sometime around 3:00 p.m., we were asked to start getting dressed for the marriage, which was set to start just before sunset. There was an uncle who knew the art of wrapping a Rajasthani turban. So, he went around and tied my turban for me and all the closest male members in my family and closest friends.

About 10% of our Baraat Guests, before getting dressed for the wedding.

After that, a band started playing Bollywood songs, alerting everyone that the marriage procession was going to begin soon. These bands, a carry-over from the British Army tradition, have about ten musicians playing different types of drums, saxophones, sousaphones, trumpets, trombones, and shehanai/Indian flutes.

As soon people got dressed, they went outside and gathered around the band. At about 4 p.m., I was led to a waiting mare/female horse, commonly referred to as just a "horse". The horse had been decorated. With bands playing popular songs loudly, I was helped on to the horse. I was paraded through the town for about half an hour, accompanied by music performed by the band and many in procession dancing to the music. While I was riding, I requested them to play the title song from the movie *Mere Mehboob*. They granted my wish. I expected them to play it once, but they played it over and over again.

MY INDIAN WEDDING

Similar to Author, Dalpat Hiran arrives for wedding on a horse.

At Sarala's home, families and friends had arrived for several days prior to the marriage. Unlike at our home, at Sarala's home, music consisted of lots of Marwari wedding songs. A mehendi/henna expert applied mehendi to Sarala and all the women and girls, and to some of the men who wanted it. At Sarala's home, they performed a Grah Shanti ceremony in which a Hindu priest read Sanskrit shlokas and performed various rituals to calm/please nine planets, that, according to Indian astrology, affects everything in your life, and to seek their blessing. They also performed Ganeshpuja, where Lord Ganesh, son of Shiva and Parvati, is requested to bless the marriage that will soon take place.

We arrived at marriage venue at about 5:00 p.m. I was still on the horse near the back end of the Baraat group. The following ceremonies of the Indian wedding took place over next few hours.

Dwarachara:

Our Baraat was formally greeted by Sarala's family members. Men were dressed in their best clothes. Many wore topis, and some wore turbans. Women and girls were in colorfully saris. All the Baraat relatives were received with a gift of coconut and some money; both sides hugged each other, and Baraatis were requested to proceed and take a seat or gather in an area close to the podium where the marriage would take place.

I was waiting on the horse, while everybody else was welcomed and proceeded to the wedding area. Some ladies from bride's side, including mother of bride, waited for me and my horse was led close to them. They were holding a brass plate with a diya/oil lamp, rice, and tilak. I bent down so they could apply the tilak. While I bent down, the mother of the bride pinched my nose. This is a mischievous way of humiliating and greeting your future damad/son-in-law. After that, she did Aarti to welcome me, applied a tilak, and offered a garland as greeting. I got off the horse and was escorted to the wedding mandap/ decorated raised platform inside a tent.

MY INDIAN WEDDING

Appointment of Priest:

Even though both Sarala and I are Jains, our wedding was performed in a traditional Vedic Hindu way by a Hindu priest, which is a common practice among Jains. The ceremony has been essentially the same for about 5,000 years. A Hindu priest read the mantras in Sanskrit. Occasionally, he explained some portions in Hindi and Marathi for the benefit of the marrying couple, parents of the bride and bridegroom, and the guests.

I had studied Sanskrit for four years, and I had done well in it. Even so, the lack of any use for six years made it difficult for me to understand what was being said. One of the great problems for me or for anyone who is not constantly exposed to Sanskrit is the compound or fused/Sandhi words. These words are made by fusing together two or more words into one. In English, a hyphen may separate two words, Indian-American for example, and a hyphen represents a short pause between two words. In Sanskrit, where the words are fused, there is no hyphen to help. On the contrary, spelling and pronunciation of fused words are altered.

Two words roll off your tongue as one word uninterrupted by a pause. If you read or hear a compound word, in order to understand it you have to mentally and quickly break the compound word back into its components to understand it, while more words are continuously being spoken. This is not so easy in a fused Sanskrit word. Another difficulty is that the words used in Sanskrit are not in commonly used today, and many of them have very difficult abstract meanings from an older time. Regardless, the 5,000-year-old tradition required the marriage ceremony to be in Sanskrit mantras/shlokas/verses. And both of us were willing to endure it and to do our best. Water, milk, KumKum/Tilak, rice, Supari/Beetleleaf, flowers, fruits, and sweets etc., were offered at various times in all the ceremonies to signify different aspects and advice of marriage.

Sarala and me, as couple, agreed to learn and practice the intent behind the ceremonies for the rest of our lives.

Ganesh Pujan:

Lord Ganesh is invited to witness the wedding and bless it. Parents from both sides participated in this puja.

Bride's Arrival:

Sarala was brought to the Mandap by her oldest maternal uncle. She was dressed in a colorful sari with a very rich gold embroidery. Her head was covered with this sari, but her full face including her forehead was visible. All brides are most beautiful at their wedding. Sarala was always so beautiful, but that day, Sarala was exceptionally so.

Bridegroom's Arrival:

A white cloth was stretched across and Sarala was requested to come and stay behind the curtain. I was escorted to the mandap and positioned before the white curtain, which was hiding Sarala. When the priest gave the announcement, the white curtain was lowered, and I could see, ceremoniously, my bride for the first time. According to old tradition, which is still often practiced, the groom sees his wife for the first time when this curtain is lowered. Everyone pretended that this is what was happening. There was a loud applause when the curtain was lowered, and everyone saw the joy in the bride and groom's faces at their "first" sight of each other. Water and sweets were offered by Sarala to me to welcome me. The diya/oil lamp was lit by both of us.

Exchange of Garlands:

The bride and groom offer a floral garland to each other and it is lowered over the head. No one told us "one at a time", so, when the priest said, "Mala Pehanao/please offer garland", both of us thought it was our turn, and both of us tried to put the garland around each other's neck at the same time. Before the garland could be lowered over our heads, the two garlands got tangled. The guests were shocked. I quickly disentangled the garlands and requested Sarala to proceed first. Order was restored. When I finished the ceremony by putting the garland around her neck, there was a big relief, laughter, and applause. To this day, Sarala praises me for my presence of mind, and for being "cool under crisis".

Kanyadan/Varpuja/Kanyagrahan:

In this three-step ceremony, the bride's parents offer to give away their daughter to the groom and his family. After that, the groom is worshipped. After that, the groom's family accepts the offer of the bride as the "greatest gift" they could receive. According to the Hindu custom, which is still practiced in almost 100% of the cases, the bride moves out of her parent's home where she was born and raised and moves into a totally strange groom's parent's home with a lot of strange people. She leaves her home with high hopes, and also some fear, for the future. It works out much better if the groom has his own place. But that is not so common. I did not have my own place.

Exchange of Rings and Hastamilap/Handshake:

I offered and placed a ring on Sarala's fourth finger and then she did the same. After the applause stopped, I was asked to place my right palm over her right palm, while the priest recited shlokas. This was the first time I touched her, and the sensation was incredible. Our scarves were tied together, and they would stay tied for a while until all the ceremonies were completed.

ShilaRohan/Walk on Rock:

Sarala was asked to place her right foot over a small piece of rock. With this gesture, we committed to overcome any and all obstacles together.

GaneshPujan/NavgrahPujan:

Sarala and I worshipped and sought blessings of Lord Ganesh and nine planets to bless us and remove all obstacles in our life. Science has demoted and removed one planet, resulting in eight planets including earth going around the sun. However, the Indian astrological tradition is unchanged.

Agnisthapan:

Holy fire which will be a witness of the marriage was lit.

Pradakhshina/Mangal-Phera:

Sarala and I walked around the fire four times.

Saptapadi/Saat-Phere:

Sarala and I took seven phere/rounds together around the fire and vowed to be married for life with the following vows taken in front of priest, our parents, our guests, the fire, and all the gods as witnesses.

> **First Step:** We vowed to remain together through happiness and sorrow.
>
> **Second Step:** We resolved to embrace each other's families as our own.

Third Step: We took an oath to remain faithful to each other.

Fourth Step: We promised to help each other in times of sorrow and rejoice together in times of happiness.

Fifth Step: We agreed to share each other—body and soul.

Sixth Step: We agreed to be married for our entire life with God as our witness.

Seventh Step: We agreed to become friends and partners and remain friends and partners for life.

Mangalsutr and Sindoor/Wedding Necklace and Red Powder:

I applied KumKum to her part in her hair and put her mangal sutr/wedding necklace made of black beads on her. Both marked her as a married woman.

Kansar (Holy Food):

We fed each other some sweets; our first food consumption together.

Purnahuti:

The priest announced that the wedding was completed and now, Sarala and I were husband and wife.

AkhandSaubhagyawatibhav:

A number of married women, close family, and friends put kum-kum/tilak on our foreheads and sprinkled rice on us and wished that I would have a long life and that Sarala would have me for her entire life. They whispered words of advice and best wishes.

Ashirvad:

The priest gave me and Sarala his best wishes, pronounced once more, the final time, that we were husband and wife. He turned to all the guests and asked them to sprinkle us with rice and offer their blessings. We individually visited our parents, our grandparents, our uncles and aunts, and other elders and touched their feet together. They blessed us by putting their palms over our backs and with words for a long and prosperous wedded life, filled joyously, with lots of children.

Om Shanti Mantra:

The priest concluded the wedding with blessings and a prayer for peace everywhere, and for us and our families. He pronounced peace three time. Shanti, Shanti, Shanti.

Sarala and I were married. Wow! Our parents, our friends, and all the guests were happy and relieved that the entire wedding went through without a hitch. The wedding was finally done! The ceremony took about three hours. Early in the ceremony, many guests recalled their own wedding and enjoyed the rituals. But soon, due to the monotonous repetitious sounding Sanskrit shlokas, they were anxious for it to be over. I have attended many Indian weddings myself. And I fully share their feelings. However, my long marriage, going on 55 years now, probably is due to the fact that it was arranged and also to a great extent because I took these vows seriously: I had vowed to be faithful and stay married no matter what.

After the marriage, on May 26, 1964, Sarala stayed with her parents, and I stayed in our hotel with my parents that night. The next day on May 27, 1964, a reception was planned for the evening. But a shocking news of the death of India's first Prime Minister, Jawahar Lal Nehru, from a heart attack changed that. Mr. Nehru, who fought for

MY INDIAN WEDDING

Independence of India under the guidance of Mahatma Mohandas Karamchand Gandhi, had been appointed Prime Minister by the congress party. Just before the clock struck midnight on August 14, 1947 and the morning of August 15, 1947, Mr. Nehru gave one of the most memorable speeches, welcoming a free India. He invited all Indians to work with him to create a great nation. His speech titled "Tryst with Destiny" began with these words: "Long years ago, we made a tryst with destiny, and now the time comes when we shall redeem our pledge..."

With this speech, and with all that he had done while fighting for the independence of India, and subsequently for 17 years after the independence as Prime Minister, Nehru had become synonymous with India. Even though he had been in poor health for a couple of years, he had not chosen a successor. He probably did not do so because he wanted India to have democracy where people would choose their prime minister, and to make sure that he did not start an heir appointment precedence. On that day, the shocking news made everyone feel sad and uncertain.

We still carried on with our life, but in a subdued way. Sarala and I visited a temple close by to seek blessings as a newly wedded couple. We spent most of the day together in company of young people from both sides. We again slept separately that night.

The next morning, Sarala departed with me to Bombay to begin our married life. I was planning to leave for the United States to start the fall semester at the University of Michigan soon, and she was going to continue with her senior year to complete her bachelor's degree in Pune. So, we had only a few months together before we would separate. We packed in as much fun as we could in that short period.

CHAPTER **33**

Departure for United States

IN 1964, WHEN I was planning to come to the United States, the number of persons traveling to the United States was a tiny fraction of number of people who travel now.

There were number of reasons for this. World population and the Indian population were a fraction of what they are today. The world population reached approximately one billion at the beginning of the 19th century after more than millions of years of human or human related species evolving. But only in two centuries, after the beginning of the 19th century, world population grew to approximately three billion in 1964. World population grew by about five billion in 56 years to the current population of about 8 billion.

The literacy rate in India and the world over was very low. Without high literacy and sufficient income, few people had the courage or the means of leaving their place of birth to go to a place unknown to them.

The mode of transportation for intercontinental travel started changing from ships, which took several weeks to come to the United States, to air travel which took only a day. Air travel was starting to become popular even though fares were much higher than by ship.

DEPARTURE FOR UNITED STATES

The United States immigration laws were very restrictive and discriminatory against Indians and Asians. The United States has a long history of "whites" doing things for their selfish interests. It started with the occupation of the Unites States by "whites" after driving American Indians out of their land, which they felt was given to them by God and Nature.

The world was divided into colonies controlled by the United States and a few "white" European countries for about three centuries when I was going to travel to the U.S. Because of this political control, the white population ruled, suppressed, and looked down upon non-whites everywhere, which included people from India, who come in a variety of shades of brown color.

In the late 19th century, the United States took steps to bar immigrants from Asia. The establishment of the quota system in 1924 reversed the bar, but it was reintroduced in practical terms by limiting the quotas to about hundred or so from Asian countries, until 1965, when the National Quota System was abolished.

The vast majority of immigrants who came into United States up until 1965 were from Western Europe. In 1882, Chinese were specifically barred from immigration. In a similar manner, the United States also found a way to bar Japanese, Koreans, Indians, and Filipinos. The quota system introduced in 1924 restricted the entry from Eastern and Southern Europe. Prior to 1945, Mexicans were hired when needed, and then mercilessly deported, in large numbers, when they were no longer needed. Between 1945 and 1965, after World War II, immigration became more diverse. Agricultural workers from Mexico were allowed under temporary visas. Asian exclusion was eliminated. Refugees were admitted in large numbers. However, Indians did not benefit from any of these reforms.

During 1848-1855, during the California Gold Rush, the Chinese, along with migrants from Latin America, Europe, Australia, Japan, Korea, and Asian Indians, were hired for needed labor. After the United States colonized the Philippines, Filipinos were allowed to enter the United States freely.

The passage of the Chinese Exclusion Acts barred Chinese from 1882 to 1943. Japanese were barred in 1907. The "white" British were welcome in large numbers, But Indians from the British Colony were barred in 1917. Asian immigrants were also categorized as "alien ineligible for citizenship" by law and/or court decisions. This ban was extended further in 1924, which barred immigrants from all Asian countries from the area known as the "Asia-Pacific" Triangle. Philippines were granted independence in 1944, partly to change the status of Filipinos from U.S. Nationals to aliens. After their independence, Filipino aliens were limited to a quota of 50 per year. Asian Americans were not welcomed in the United States until the end of World War II.

After World War II, many exclusion laws were reversed, and Asians became eligible for citizenship; but a quota of 100 per country, and total quota of 2,000 from the entire Asian continent did not change things much. However the reversal in these laws did prepare the United States population to start thinking of Asians in a different, more favorable manner, which may have made the passage of the immigration reform of 1965 easier.

There were many "non-quota" categories. Some Asians were able to use these categories to come after World War II. Many Asian wives, brides, fiancées, and children of military veterans and/or United States citizens entered the United States in this manner. All previous Asian immigration was primarily male. After World War II, Asian women arrived in larger numbers, making up a great deficit in the Asian women population at that time.

DEPARTURE FOR UNITED STATES

After World War II, because of the United States' emergence as the biggest superpower, its behavior needed to change. It became more involved in international affairs and had to increase the number of refugees admitted. Refugees arrived first from Germany and German-occupied Europe, the majority of whom were Jewish. There were about thirty million displaced Europeans. United States took about 400,000 of them as refugees. The start of the cold war between the United States and Soviet Union immediately after World War II led to admission of those who claimed to be political refugees from communism. However, this generous act by the United States primarily benefited Eastern Europeans. Subsequently, another wave of Hungarian and Cuban refugee followed. Few Asians, and even fewer Indians benefitted from this change in attitude until 1965.

Bias in favor of Western European is obvious from the fact that, in 1952, quota allotments were as follows: Great Britain 65,000; Germany 26,000; Ireland 18,000; other Europeans 40,000. In contrast, the quota for all Asians was 2,990, and the quota for all Africans was 1,400. Not only that, but over 400,000 Europeans came as refugees or displaced persons.

This was the state of immigration after more than 50 years after Swami Vivekanand made Americans aware of the intellectual contribution that Indians could make. India had a lot to offer in cultural and spiritual understanding, and in mathematics and sciences. My visa was therefore restricted to the area of study. There was a high likelihood that I would be allowed an additional stay for up to 18 months to obtain practical training. Therefore, in less than 3 years, I would be required to leave the United States because of the brown color of my skin and my Asian Indian national origin.

The 1965 Immigration and Nationality Act changed discrimination against Indians and all Asians. The National Quota system established in 1924, which limited the number of Indians admitted to the United

States to 100 per year was abolished. Now the number was 20,000 per year. In addition, there was greater emphasis on family reunification with the 1965 Act. Immediate family members, including spouses, unmarried minor children, and parents of these new immigrants from India and Asia received unlimited non-quota status.

The law was partly passed under an erroneous assumption and/or explanation by senators and congressmen that the old proportion of national origin of immigrants would not be severely disrupted. In 1965, the Asian population was small, and these politicians expected that only a small number of Asians would apply. Upon realizing their error, further amendments did restrict the number of Asians allowed to come in. But these restrictions did not change the new reality that United States was now open to everyone around the world, regardless of the color of skin, amount of education, level of command of English, religion, or the amount of wealth of the immigrant. Everyone was welcome. My wife and I would be welcome as well. Fortunately for me, this change came at just the right time.

With very few Indians going abroad, my planned travel was a big deal. Because of this, Babuji's close friend Rajkrishna threw a party at the Mafatlal swimming bath and boat pool in South Bombay near Charni Road. About fifty persons were invited for the dinner party, in my honor, which took place on the deck next to the swimming pool.

I would advise anyone accepting a farewell party to be sure they are definitely going to leave. After the party, it is too late! In my case, when the party invitation was accepted, I knew definitively that Baiji had finally accepted my decision to go to the United States. So now, I and everyone connected with me started announcing the date of my upcoming departure in night of August 16th, and the early morning of August 17, 1964.

As the word spread during the last few weeks of my stay in India,

DEPARTURE FOR UNITED STATES

everyone whom I met expressed admiration for me, my accomplishments, and my luck. They gave me their best wishes and blessings. All the relatives who could travel to Bombay to see me go off to the United States made their travel arrangements and started arriving a few days before my departure.

Packing bags for this travel was a mammoth undertaking. There were many considerations. The biggest consideration was that everything that I was going to carry had to be kept within free baggage allowance. Excess baggage fee was prohibitively expensive. Anything that was shipped had to be shipped by ship to keep the cost reasonable. Therefore, this separate package would not arrive for several months.

There was great uncertainty about the availability of vegetarian foods in the U.S. Therefore, Baiji insisted that I take oil-based pickles which had been sealed in a can with molten tin metal. There was a grave risk. If the seal did not hold, all of your clothes would be stained red with red pepper, and yellow with turmeric which are used as spices in the oil base of the pickles. I expressed this concern, but Baiji insisted and I went along.

There was also uncertainty about the weather. For anyone who grew up in Bombay where temperatures rarely go below 60 degrees F, the biggest concern was how cold it gets in the United States. I wanted to make sure I would not freeze. So, I wore my thermal underwear and my suite over it to keep warm. I did that even though I would be starting my trip from Bombay, which is warm throughout the year. I did not realize that I was arriving in New York in August, which is the hottest month for New York with high temperature in the 90s F.

We carried my bags several flights up and down, several times, in the process of packing, to a local grocer. We used his traditional weighing scale, the kind you see in Mughal-e-Azam with standardized steel weights on one side and merchandise on the other.

GROWING UP IN MUMBAI, INDIA IN 1940S, '50S AND '60S

Group Picture before Departing for Airport.
Standing from Left: Sheela, Virendra, Dilip, Sushil, Prakash, Anil.
Sitting from left: Urmila, Babuji, Preeti, Baiji, Author in Suit, Author's wife Sarala, Atul, Nirmala

To keep my free baggage weight under limit, I added a lot of heavy books and other heavy items to my previously weighed and tagged handbag, even though I did not need them during travel.

It was getting close to 7 p.m. The flight was to leave at 1 a.m. the next morning. Even though we still had six hours to go, everybody started going into a typical panicky travel mode. Everybody was doing something to help rush things, often only to slow them with their distraction. But this was the most fun part of travel in India. Even though I was the one and the only person traveling, about forty guests that had arrived from all over India to see me off were doing something "useful" for my trip.

It was time to get all the members of my family and all the guests, that had traveled from all over India, to the airport to see me off. So this whole group got into taxis, tangas, rickshaws, bikes, trains, and buses, and make it to airport by 9 p.m. We made sure that everybody

DEPARTURE FOR UNITED STATES

who had taken so much trouble to travel from all over India to see me off were all able to make it to the airport.

About 10 p.m., after talking briefly with everyone and telling everyone that I was going to miss them, I waived them goodbye and walked inside the terminal toward the ticketing area.

L to R : Sushil, Santosh, Dilip, Prakash, Viru, Anil Garlanding

Anil garlanding Author just before take-off. From Left to Right: Sushil, Author, Dilip, Prakash, Virendra.

A few people, including Sarala, were allowed to join me into the ticketing area. Some may have bribed or pleaded successfully to get this exceptional access. After my bags were checked in and I received my boarding pass, I said my final goodbye to those who made it to the ticketing area. Sarala stepped aside. I walked up to her, exchanged a few words, shook her hand and said goodbye to her and everyone again. I did not kiss her. You do not display your love, in public, in India.

GROWING UP IN MUMBAI, INDIA IN 1940S, '50S AND '60S

In a few hours, I would be leaving India for a strange land, to live among strange people, without my wife, my friends, my family, and my countrymen. I was sad one moment and I was happy the next. I was a young man with a long life ahead of me. I knew I had made the right decision. I proceeded through immigration and the customs check and then to the boarding area. After a long wait, I boarded the plane. The plane took off. Shortly after, the lights were dimmed. It was 2 a.m. and I dozed off.

CHAPTER **34**

Superman, Loise Lane, and I

SUPERMAN, A COMIC book superhero character was published starting in 1938 by DC Comics. It was made into one of the biggest blockbuster movies in 1978 titled *Superman*.

Superman, born in another Galaxy, lands on earth as a baby. He is adopted and raised by a rural family. He is given a name of Clark Kent. Clark works for a newspaper as a reporter where he meets Lois Lane, who is also a reporter. He falls in love with her. One day, Lois Lane is going on an assignment in a helicopter. The helicopter malfunctions at takeoff and ends up crashing into the parapet of the terrace roof of a 50-story building where the Daily Post's office was located. The helicopter is dangling. Lois lane, who was not wearing seatbelts, slides off, but keeps herself from falling 50 stories by hanging onto a part of helicopter frame. A crowd gathers on the street below. The police pull in and tries to figure out how to save her from certain death if she loses her grip. Time is quickly running out. Clark becomes aware of the accident. He must quickly become Superman to try to save her. Lois is tired and loses her grip and falls. Superman is able to stop her midway through this fall and saves her life.

Superman was produced in 1978, distributed by Warner Brothers, and starred Christopher Reeve as the superhero and Margot Kidder as Lois

GROWING UP IN MUMBAI, INDIA IN 1940S, '50S AND '60S

Lane. It was the most expensive movie at the time with a budget of over 50 million; and it was second highest grossing movie of the year with $300 million gross income. It remained one of the most popular movies for a long time, partly because of the scene I described above. You will find out shortly how this scene is connected to me.

I was often sick as a child. I remember lying in bed many times each year with some kind of pain or fever, with Baiji staying up all night to comfort me. The reasons were many— improper nutrition, bad hygiene, and poor medical care, along with dirty air and contaminated water. I remember being sick one time on the night of Diwali and the morning of Saal Mubarak/Hindu New Year. This was a very special day of joy. On this day, all the children visited many of our closest friends in the Kavarana Mahal. We received gifts of sweets and salty snacks. But because of my sickness, I was not allowed to go that day. In India, rest is often the most commonly tried remedy to cure illness. Some colored medicine prepared by a doctor's compounder was given, which made you feel better; sometimes, even if it was not effective.

During my engineering college years, I used to complain about a cough. My doctor ordered some x-ray screening that confirmed an infection in my lungs. The medicine prescribed by my family doctor was not having much beneficial effect, so Babuji and my family doctor decided that I needed to see a foreign trained lung specialist. He charged 15 rupees for one visit and prescribed an injection that I had to take for six months. My family doctor, also a family friend, Dr. Gandhi, whom I had been seeing all along, administered these injections, which I received, several times a week, on my way back from college.

One day I stopped by Doctor Gandhi's clinic to receive my shot. He was suspicious when he touched me to give me a shot. So, he took my temperature, and informed me that I had 104-degree fever. He

SUPERMAN, LOISE LANE, AND I

gave me an additional shot and some medicine. He instructed me to go home and lie down and rest. I told him that my exams were only a couple of days away; that I had to prepare and that there was no way I was going to lie down and rest. He said that because of my fever and pneumonia, I should forget about my exams. I went home by train and informed Baiji. Baiji informed Babuji when he came home about my medical problem, and that Dr. Gandhi recommended rest and skipping my college exams. Babuji said that I was well prepared and that I didn't need to worry. That I should go ahead and take the exam, and that I would be fine. I took the exam. He was right. I stood first in my class. I broke records in many subjects and in overall marks.

In addition to my medical problems, I had too many close dangerous calls.

More than once, at Shivaji Park Beach, there were some retaining walls. I had jumped off these walls onto sand or concrete sidewalk or slab, only to end up in great pain because I did not realize how tall the wall was. When the pain was great, I slept in Baiji's lap. She gave me a sleeping medicine called Gulkanda—a sweet preserve, made by crushing rose petals, which served as a sleeping pill.

Another time, I remember running across Lady Jamshedji Road to catch up with my friends, who were ahead of me and had already crossed the road. Before I knew it, I was hit by a car and thrown onto the sidewalk. I got up, limped home, got some rest. Thank God there was no permanent damage.

Local gangs were everywhere because of high unemployment. They collected money and used some of it for social work, or some religious activity. Diwali celebration was one such occasion.

They had built a huge kandil, a paper lantern framed by wooden sticks, which was about 15 feet x 15 feet. This huge Kandil was hung between our building and one across Lady Jamshedji Road, about 100 feet apart, at the 3rd floor level.

Diwali is celebrated with firecrackers. I was doing the same. I had a burning sparkler in my hand. I suddenly had the urge to hurl this sparkler in the air, which I acted upon immediately. Unfortunately, to my shock, the still burning sparkler landed on the huge Kandil. Within a few seconds there was an empty space where Kandil was. In a few minutes, the gang noticed the missing Kandil. They were very upset. The gang had put it up with so much love, effort, and money. They had to find the culprit and punish him.

I have no idea how, but they figured out that the culprit had to be from our building, and the person doing so must be from my 4th floor, the only floor above the level of Kandil. So, they came to my floor and confronted all the kids on my floor. No one admitted it. I was also questioned, and I did not admit it either. They left with a threat that as soon as they found out, they were going to beat the hell out of the person. They did not find out. I thanked Lord Rama, whose triumphant return to Ayodhya was being celebrated.

Another time, I was going out with my friends. Against my desire, my youngest brother, Anil, wanted to tag along. So, I ran downstairs to try to lose him. In the process, I ended up missing couple of steps. I fell and hurt myself. I was in pain for a while.

Bombay has about 90 inches of rain, concentrated during four months of the monsoon/rain season from June to September. For about four months, there was rain almost every day. Most of the time when it rained, it rained very hard. A slow drizzle is not that common. And when it rained, there were water puddles everywhere—on the road, on the sidewalk, on the empty land. In our building, we had puddles in exterior corridors and the stair area. I loved those puddles. The reason was that these areas were finished with smooth

SUPERMAN, LOISE LANE, AND I

concrete; when there was a puddle, you could run on the dry part barefoot, pick up speed, and glide on the puddle part. After gliding across the puddle, you would slow down and stop. One time, I picked up too much speed and lost my balance while gliding; I fell and hit back of my head on the hard concrete. I will never forget the sharp pain. Even so, the thrill was too much, so we continued to slide and hurt ourselves often. As long as I did not bleed badly, I didn't care.

Now that I have prepared you somewhat, I want to tell you about one more "accident". This accident involves me falling from a fifth floor to the concrete sidewalk on street level. Since I am writing, obviously I survived. But you definitely want to know more, right?

Bombay is an island city surrounded by marshland. Hot humid weather and rain is perfect for mildew and moss. These make the concrete turn gray, black, and green.

Kavarana Mahal, like most buildings in Bombay, are topped with a flat terrace. These flat terraces provided additional recreational space which is in very short supply in Bombay. The flat terrace was protected with a brick parapet wall.

When I was 11 and in 6th grade, I was running and playing hide and seek with my friends.

To hide in a difficult place, I came up with a plan. There were two water tanks. One on the terrace of first building where I lived, and another one on the terrace of the second Kavarana building behind the first building. The tank on the second building was one story lower. My plan was to get on top of the first tank, jump down to the lower second tank, and then hide under the second tank. In order to get on top of the first tank, I jumped up to the top of a parapet wall near the first tank, which was 3' tall. Then, I ran a few feet on top of the 8" wide parapet wall so that I could jump up to the top of the first tank. I had done these many times before.

There was green mildew/moss on top of the wall. This type of moss is very slippery. As I was running on the wall to get ready to jump to the top of the tank, I slipped and fell off the side of the building. I was dropping, and I was on my way to splash on the concrete sidewalk within a few seconds, five floors below.

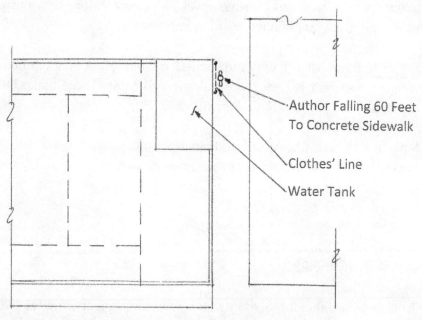

Author tried to get on top of water tank.

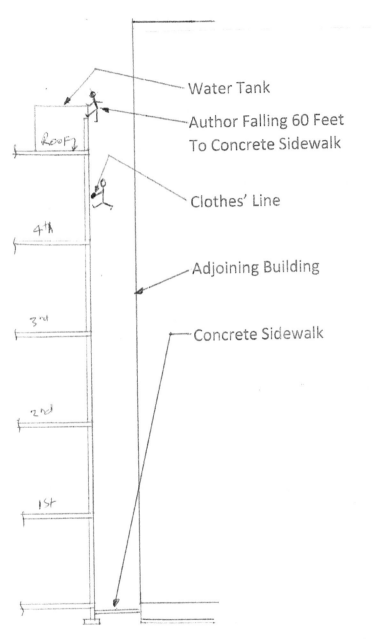

Author slipped and fell. Author was lucky to catch clothes' line and survive unhurt.

GROWING UP IN MUMBAI, INDIA IN 1940S, '50S AND '60S

But as destiny would have it, I found a rope stretched for drying clothes on the fourth floor. I grabbed it. The rope did not break. I quickly climbed back onto the fourth floor parapet wall. I was safe. I was probably in shock, but I do not recall any fear. I did realize that I had a close call. But I did not think much about what might have happened. Somehow, we all have optimism that things are going to be all right and that helps us to go on. It did not seem all that unexpected that I had grabbed the rope and survived. Why would I not expect to?

I climbed the wall which was part of Chandu's apartment. I entered his apartment. I ran through the apartment and made sure no one saw me. I did not tell anyone what had happened for a while. Then after I had recovered from the shock, and after I felt that anyone listening to my story would not go hysterical, I began telling them. Amazingly, everyone's reaction was so mild.

A comparison between a scene from Superman: The movie and my situation might be in order.

Lois Lane was doing her job, so her accident was not her fault. I was careless and was walking on top of a wall that was not intended for walking.

She was falling from a 50-story high building, and I was falling from 5-story high building. Gravity could not care less. Both of us would be dead if we met the ground.

There were hundreds of people and some helpless police trying and praying for her safety. I was alone.

When she was trying to hang on, and when she was rescued, there was screaming and loud music in the background. For me everything was quiet.

SUPERMAN, LOISE LANE, AND I

Moviegoers lined up to see her being saved. No one heard of my story for a while. And when they did, there was not much of a reaction.

Superman saved her. I was saved by my cool head, presence of mind, and my dexterity in grabbing the rope to save myself.

One common thing—both of us needed a lot of luck, and we both were extremely lucky.

Superman actor Christopher Reed later had a horse riding accident which paralyzed him, and he died fighting to recover valiantly in 2004 at age 52. Lois Lane actress Margot Kidder went broke, suffered mental illness, but never gave up, and died in 2018 at age 70. I suffered a mild heart attack at age 74 and I am writing this book at age 77.

I suddenly woke up. The stewardess was announcing that we were beginning our descent to land at New York John F. Kennedy International Terminal. I just realized that I had been dozing on and off throughout the trip. Amazingly, I was seeing a dream that continued from dream to dream in an almost continuous story of my life to this point. Amazing! Weird too!

We needed to buckle up and get our customs and immigration papers completed now. We began our final descent to land. The pilot reminded that we needed to tighten our seat belts and bring our seats to the upright position. Stewardesses were walking up and down the aisle making sure we had done so. We were about to land. I was going to enter the United States for the first time, and begin a three-year trip, which eventually would turn into a lifelong stay.

CHAPTER 35

My Arrival in New York

I WAS SITTING in the plane, which was going to land shortly in New York, looking down, at clouds below. It was August 17, 1964. After about 9 months of dealing with bureaucratic red tape, I had obtained a student visa to go to the University of Michigan to study for a master's degree in Civil Engineering. My boss at my last job in India had told me that if I came back to work for them after completing my studies, I would be getting a more responsible job with a raise to Rs. 800 per month from my last salary of Rs 350 per month. So, when I made the decision to go to the U.S., it was a no-brainer. I would easily recover cost of my travel and education by working for 18 months, I would be allowed to work on a student visa to obtain practical training. And I would be making more money for the rest of my life. There was no net cost or risk to me. So, I was looking forward to my new life for next three years in the U.S. as the plane was making its descent.

The Air India plane arrived at New York City's John F. Kennedy International Airport at about 3:30 p.m. on August 17, 1964, right on schedule. I was expecting to meet my classmate and friend, Kanti, who was in New York City for the summer. He had preceded me in coming to the U.S. by a year. I had mailed him my itinerary. I had also written to an organization called Friends of Overseas Students, whose

MY ARRIVAL IN NEW YORK

student volunteers welcomed new international students to the U.S. and helped them at the airport to get started.

When I got through with immigration and customs, and I walked into the arrival baggage claim area, I knew that I had overcome all the obstacles. I was INDEED IN THE UNITED STATES!

I claimed my baggage and looked around to spot my friend, whom I expected to be waiting for my arrival. But he was nowhere in sight. I tried his home phone number, but there was no answer. That was the only number I had. I enquired about the organization that was going to welcome me. Someone guided me to a couple of young men who were standing close by. I introduced myself. They checked their list. They found my name and welcomed me. I informed them that I was expecting my friend to receive me at the airport. They also tried his phone number but had no luck. They asked me to wait for about an hour to see if he showed up. They said something about New York traffic. So I waited.

After this long wait, I decided that one of my closest friends had let me down. He was not coming. So, I talked to the young men who were helping me. After they found out that I had only eight dollars in my pocket, they recommended that I spend the night in a seedy hotel in Times Square. They gave me information on which bus to get on, and also about the transfer to another bus which would get me close enough to the hotel that I could walk.

Being from Bombay and having used public transportation, and even though I was very tired, I did not panic. I made it to the hotel, which had a large neon sign, at about 7 p.m. It was summer and it was still daylight. I checked into the hotel. The room had a vibrator bed. I set my bags in a corner. In a few minutes, a man knocked on door. He asked me if I was interested in a woman. I told him I was married. His expression conveyed a surprise at my reply. He said,

"So?" I told him, "No thank you." He departed. I closed the door. I sat on the bed, started thinking about how I was going to manage my stay in New York. I had no plan B since my friend Kanti was going to take care of it all, so I thought. I was exhausted, frustrated, and unhappy. I sat on the bed, disappointed: *why did my close friend let me down?*

In Indian culture, particularly at that time, a guest was supposed to be treated like God. So, when an out of town visitor came by train, you were expected to take care of all his comforts and needs, including transportation, meals, and lodging, from the time the train arrived at the station with him, to the time the train takes off with him. This did not depend on whether his stay was going to be short or long. This also did not depend on how inconvenient all these arrangements may be for the host. If you had to take a few hours off or even a day off to do so, there was no hesitation to do it. Your boss understood and you took time off for this type of activity, many times without advance notice.

I was therefore shocked that after my first 24-hour long trip covering 8,000 miles, my close friend would not even show up at the airport. It was particularly disappointing because I had contacted him long in advance, and he had not expressed his inability to help me. The only justification I could think of was that maybe the mail service never delivered my letter to him. He had not confirmed that he was going to be at the airport. But based on my habit of Indian culture, he did not have to, and I could and did assume that he would be at the airport. I was very disappointed. I had to quickly decide how I was going to change my plans for this unexpected turn of events.

I sat on the bed, thinking, worrying. I had no idea. I went to sleep, still in my suit and my thermal underwear that I had been wearing for over 24 hours.

MY ARRIVAL IN NEW YORK

Suddenly there was a hard knock on the door. I was so sleepy; I did not want to deal with that prostitute agent. But I dragged myself to the door. I opened the door. Lo and behold, it was Kanti at the door!

I was relieved and extremely happy and surprised to see him at the door. I welcomed him in. I asked him to sit down. He refused. He said he had come to take me to his apartment. I told him that I had already paid for the night. He said he was not taking "no" for an answer. I was going to leave with him right then and there. I was so happy. I was ready to give up on a friend who had miraculously located me, and who was not going to tolerate any further separation until I left New York.

Kanti was in New York doing summer jobs and lived in an apartment in Queens. He had received my letter, and he had planned to pick me up at the airport. He was paid hourly. He could not afford to lose his pay for the hours that he would not be working if he had tried to arrive at the airport at 3:30 p.m. In India, most jobs, including mine, paid a salary, and I would not have lost any money if I had taken off in a similar circumstance, so I did not think of this possibility.

I was extremely happy to discover that my friend had not lost his Indian culture, as I had incorrectly started assuming. I knew then that western culture may corrupt an Indian, but it wouldn't happen overnight, and maybe never.

I stayed with him for three days. Suddenly I remembered that in all the excitement, I had not informed my parents of my safe arrival. So, I sent a Western Union telegram of seven words which cost me $7 (current value approximately $100).

During the day, I visited the New York World Fair in 1964 by subway system. In the evening, Kanti prepared some kind of meal. We reminisced about our college life, updated each other about our common friends who were in India or who had come to the United States. He was happy to tell me about the differences in Indian and American life. I tried to learn as much as possible about these differences to prepare myself for my new life.

CHAPTER 36

Arrival in University of Michigan

I TOOK A greyhound bus from New York to Ann Arbor, Michigan. The 600-mile trip took about 12 hours. I arrived at the International Center at the University of Michigan. They made arrangements for me to stay with an Indian student for the night.

The next day they paired me with another new Indian student who could be a potential room partner. I would have never considered rooming with a Muslim in India. But the difference of religion melted away so easily within a few days of my arrival in the USA. There were so few Indians. Mohammed was from Bombay and spoke Hindi/Urdu. This commonality was enough for us to feel like friends almost immediately. Mohammed and I looked at a few apartments and also got to know each other a little bit and decided that we could room together. We found a house on Haven Street, about a mile from the Civil Engineering Building, where most of my classes would be held. The landlord, who lived on the first floor, was renting three bedrooms upstairs, which shared a common bathroom. Mohammed and I shared one room and paid $40 each for our room. The other two bedrooms were already rented to two white American students, each with their own room.

I found out that our landlord worked as a janitor. When I looked out the window, I saw a car. The car belonged to the landlord. In India, I never saw a bhangi/janitor who did the same menial work of cleaning, as our landlord, to have any wealth at all. The janitors in India were always extremely poor. They would generally live in a slum. They would walk to work, because they often could not afford bus fare. I was amazed that a person doing the same kind of work in the USA could own a home, a car, and be making additional money from rent as our landlord. I knew that day that I had come to the land of opportunity. I realized that in the USA, no work is small or big, if you do it with pride, and do it as well as you can. I was expected to treat my janitor/landlord with same respect I would show to everyone else, and I learned to do so.

I also quickly learned that my roommate and my two American neighbors ate beef, drank alcohol, and smoked. I would have never associated with such persons, let alone room with one. But I was in a foreign country. I would have to compromise and learn to adapt. Soon we all became good friends. My partner Mohamed was from Bombay, spoke Urdu, which is nearly same as Hindi. So we had a lot in common. Both of us cooked separate meals in separate pots. He became more vegetarian for convenience. I stayed vegetarian and did not compromise. However, I did tolerate him cooking beef and consuming it with relish in my presence.

Thus began my stay in the United States, which was to be completed by 1967. 54 years later, I am still in the United States writing about my childhood in India.

About the Author

GROWING UP IN MUMBAI, INDIA IN 1940S, '50S AND '60S

Santosh Kothari, Author, is a Registered Professional Structural Engineer and a commercial real estate developer and investor. He arrived in United States of America from India in 1964 at age 20 with $8 in cash and a dream to learn and prosper. He has exceeded his wildest dreams.

He graduated from University of Michigan with a Master of Science in Civil Engineering in 1965 at the top of his class in 10 months. He quickly progressed in his employments to a title of Chief Engineer at age 34. He quit this lucrative job at age 37 to start his own consulting practice and designed close to 1000 projects valued between one million and ten million US dollars each In United States and some countries in North America, South America, Europe and Asia- really in all of world's continents except Africa, Antarctica, and Australia. He became a multi- millionaire when he added commercial real estate development and ownership to his business.

He has been active in his temple/church known as Jain Society Of Greater Atlanta and has held many positions including President and CEO. He was doing all this with enthusiastic help from his beautiful and smart(er) wife while raising two sons and one daughter, all of whom graduated with two college degrees each without any student loan.

At age 78, author is physically active walking 4-6 miles on hilly roads every day. He also does a few hours of light aerobics group exercise, light yoga and light meditation. He enjoys reading magazines such as Economist and Scientific American, and books on science, politics, religion and variety of other fields. He loves singing Bollywood Hindi songs and has a youtube.com channel that can be searched with phrase "Remix Santosh Kothari" and title of song if selected.

When author first thought of writing this book, India was going thru exponential growth in population and prosperity. Author wanted to

ABOUT THE AUTHOR

preserve and share the story of his life in India which was so starkly from his life in his adopted country, United States. India, where he was born, had also changed greatly over 57 years since he left. Covid-19 Coronavirus has produced conditions in India and United States which remind him of the difficult times he grew up in.

Author hopes that reader will enjoy going back to, or learning for the first time, life in India in 1940s, 50s and 60s and discover differences and similarities to what he or she has experienced in his or her life. Author hopes that reader will have many moments of discoveries and accompanying joy. Author is presenting a personal story, but he has taken liberty to digress and provide many details and stories of general nature connected to his life.

Story behind "Growing Up In Mumbai, India in 1940s, 50s and 60s"

After my first edition of this book was printed, many people asked me questions on how I became an American author. The answer I gave them probably is not very inspirational. Even so, here it goes.

After I started my third job and had worked for a couple of years, I had learnt a new technique of solving engineering problems using nomograms in 1967. I don't think almost anyone born in 21st century would even know what nomogram is. In order to know what nomogram is, you first need to know what logarithm is. Most of you who finished high school using calculators generally would have had no need to know what logarithm is. Nomogram is a technique which uses properly distanced lines that are marked in logarithmic scale (look it up in Wikipedia or Google). You can solve complex mathematical formulae by drawing erasable lines on nomograms. I developed some nomograms to solve calculations related to reinforced concrete and printed a paper in Journal of American Concrete Institute in 1966.

My daughter got married in 1995. Happy, exhausted and relieved after the marriage was completed and bills had been paid, I thought an

STORY BEHIND "GROWING UP IN MUMBAI, INDIA IN 1940S, 50S AND 60S"

entertaining and educational manual on Indian Marriage conducted in United States might be a good one to write. I started on the book. After I wrote a few pages, information on these few pages was accidentally revealed to my wife. She did not approve of some portion. That discouraged me enough to put a quick end to my budding writing career.

I became active in my church named Jain Society of Greater Atlanta from 1989, after we moved to Atlanta from Chicago. My new Jain friends encouraged me to attend many lectures given by visiting scholars to our church. I also got interested in reading books on Jainism. In 1996, I was elected to become President of our society. The need to speak publicly in front of a Jain audience increased my desire for more knowledge in Jainism. Towards the end of my one year term, I felt that I had acquired sufficient knowledge to write a small book on Jainism. And I did so. I printed it privately and circulated it to a few friends and a few scholars that stayed with my family while visiting our church. My friends and these scholars paid complements to me for my effort. Highly respected our Gurushri Acharya Chitrabhanuji also complemented me. But he went further and invited me to join him. He said if I choose to do so, then within one year, he expected me to be speaking publicly with him on his lecture tours. One Shramaniji also requested a copy of my book. She explained that she could use it in her presentation to American born Jain kids who needed to be taught in English since they were not fluent in Indian languages.

I succeeded in engineering profession. I rose to a title of Chief Engineer in 1974, in just nine years after start of my profession. I started my own consulting engineering company in 1976. While running the engineering company, I also was successfully involved in developing my own commercial projects. I generally did a project at the rate of one project every three years. First year was devoted to acquisition. Second year was devoted to planning. Third year was used for actual construction. All projects were successful.

GROWING UP IN MUMBAI, INDIA IN 1940S, '50S AND '60S

On my 70th birthday in 2012, I looked back at my success and felt good. I felt like sharing story of my life with people who may not know me. My plan was to write a glowing book about my success in life. It became apparent very quickly to me that this was a hazardous project. There are bound to be parts that might offend people close to me, particularly my wife. So I dropped the idea.

After my 75th birthday in 2017, I came up with an idea of writing about the least controversial part of my life, my childhood. This was a story that I could tell safely. It took me three years to write and publish it. I enjoyed writing the book. I hope you enjoy reading it.

Reviews

I wrote the book mainly for the pleasure of writing. I was surprised at the variety of reasons, conveyed to me in conversations and by written reviews- some are reproduced below, why readers chose this book to read. For example:

1. I am an Indian-American parent or a grand parent. I want my children or grandchildren to know the difficulties that we have faced, and how we worked hard and rose above our circumstances to provide them with the best opportunity we could provide.

2. Your book has so much information about India. I am going to keep this book on my bookshelf. I am hoping that my children will be curious and pick it up when they feel like reading it to know more about their heritage.

3. It is refreshing to see a book about someone who is not a celebrity, and yet has so many interesting personal stories to tell. I was really surprised that this book could be so interesting. This book can serve as a personal historical book of the time not only of the author but my childhood as well.

4. I don't know much about India. After reading your book, I feel I know enough about India to hold a good conversation with my Indian-American colleagues.

Here are the actual words from some readers.

GROWING UP IN MUMBAI, INDIA IN 1940S, '50S AND '60S

Sunita J. :

Endearing personal memoir entwined with historical events in India and the USA. A great read for all generations.

Sejal M.:

I am not able to keep the book down and sleep, as I am curious to read all the details of Indian history and family history melting together. The book gives detailed explanation of all the traditions, history and the value systems. It is commendable how the author has incorporated pictures and his story in a book that is easy to follow. Author has done a great job. I will cherish this book. I am hoping that my kids will read it someday and see India from author's eyes. I Can't wait to read all of it!

Grace M. :

A fantastic read on the journey of the author and the journey of a nation of his birth, India. A vivid portrayal of growing up in India and migrating to USA. It is well woven into context of the cultures of India and America. I loved it!

Tarneka W. :

Outstanding "read" as the author takes you on a personal journey of life and Indian culture. This book is an engaging trip through the author's life. The words from the book jumped off the pages, as if I was actually there. By reading this book, I learned a lot about the Indian culture and how much support is given by each family member. If you are looking for a thrilling ride through the Indian culture- the highs and the lows, then you should not hesitate to pick up a copy of Growing up in Mumbai, India.

REVIEWS

Neela D. :

I ordered the book because I want to read it and I want my daughters to read it.

Divya P. :

I was very impressed with this well written book. As an example, I really enjoyed reading how author met his wife, details of his marriage ceremony and his explanation of what these Indian marriage ceremonies mean.

Sarala K. :

I don't read a lot of books. But this one I finished in no time. I am amazed at the amount of research and information contained in this memoir about such diverse subjects such as Independence of India on one hand and history of all the places visited by author in a trip with friends.

Anonymous:

I read this book from cover to cover as soon as I got it. I enjoyed it so much that I have already shared it with my friends. Now I am going to share it with my neighbors!

CPSIA information can be obtained
at www.ICGtesting.com
Printed in the USA
FSHW012009170821
83924FS